Nanostructured Materials
and Nanotechnology V

Nanostructured Materials and Nanotechnology V

A Collection of Papers Presented at the 35th International Conference on Advanced Ceramics and Composites
January 23–28, 2011
Daytona Beach, Florida

Edited by
Sanjay Mathur
Suprakas Sinha Ray

Volume Editors
Sujanto Widjaja
Dileep Singh

The American Ceramic Society

A John Wiley & Sons, Inc., Publication

Published by John Wiley & Sons, Inc., Hoboken, New Jersey.
Published simultaneously in Canada.

For general information on our other products and services or for technical support, please contact our Customer Care Department within the United States at (800) 762-2974, outside the United States at (317) 572-3993 or fax (317) 572-4002.

Wiley also publishes its books in a variety of electronic formats. Some content that appears in print may not be available in electronic formats. For more information about Wiley products, visit our web site at www.wiley.com.

Library of Congress Cataloging-in-Publication Data is available.

ISBN 978-1-118-05992-0

oBook ISBN: 978-1-118-09536-2
ePDF ISBN: 978-1-118-17267-4

ISSN: 0196-6219

Printed in the United States of America.

10 9 8 7 6 5 4 3 2 1

Contents

Preface

The 5th International Symposium on Nanostructured Materials and Nanotechnology was held during the 35th International Conference and Exposition on Advanced Ceramics and Composites, in Daytona Beach, Florida during January 24–29, 2010. This symposium provided, for the fifth consecutive year, an international forum for scientists, engineers, and technologists to discuss new developments in the field of nanotechnology. This year's symposium had a special focus on the large-scale integration of functional nanostructures and challenges related to the fabrication of nano-devices. The symposium covered a broad perspective including synthesis, processing, modeling and structure-property correlations in nanomaterials and nanocomposites. Over 80 contributions (invited talks, oral presentations, and posters) were presented by participants from universities, research institutions, and industry, which offered interdisciplinary discussions indicating strong scientific and technological interest in the field of nanostructured systems.

This issue contains 17 invited and contributed papers peer-reviewed using The American Ceramic Society review process and covering various aspects and the latest developments related to processing of nanoscaled materials including carbon nanotubes-based nanocomposites, nanowire-based sensors and electrode materials for lithium ion batteries, photocatalysts, self-assembly of nanostructures, functional nanostructures for cell tracking and heterostructures.

The editors wish to extend their gratitude and appreciation to all the authors for their cooperation and contributions, to all the participants and session chairs for their time and efforts, and to all the reviewers for their valuable comments and suggestions. Financial support from the Engineering Ceramic Division of The American Ceramic Society is gratefully acknowledged. The invaluable assistance of the ACerS's staff of the meetings and publication departments, instrumental in the success of the symposium, is gratefully acknowledged.

We believe that this issue will serve as a useful reference for the researchers and technologists interested in science and technology of nanostructured materials and devices.

SANJAY MATHUR
University of Cologne
Cologne, Germany

SUPRAKAS SINHA RAY
National Centre for Nano Structured Materials
CSIR, Pretoria, South Africa

Introduction

This CESP issue represents papers that were submitted and approved for the proceedings of the 35th International Conference on Advanced Ceramics and Composites (ICACC), held January 23-28, 2011 in Daytona Beach, Florida. ICACC is the most prominent international meeting in the area of advanced structural, functional, and nanoscopic ceramics, composites, and other emerging ceramic materials and technologies. This prestigious conference has been organized by The American Ceramic Society's (ACerS) Engineering Ceramics Division (ECD) since 1977.

The conference was organized into the following symposia and focused sessions:

Symposium 1	Mechanical Behavior and Performance of Ceramics and Composites
Symposium 2	Advanced Ceramic Coatings for Structural, Environmental, and Functional Applications
Symposium 3	8th International Symposium on Solid Oxide Fuel Cells (SOFC): Materials, Science, and Technology
Symposium 4	Armor Ceramics
Symposium 5	Next Generation Bioceramics
Symposium 6	International Symposium on Ceramics for Electric Energy Generation, Storage, and Distribution
Symposium 7	5th International Symposium on Nanostructured Materials and Nanocomposites: Development and Applications
Symposium 8	5th International Symposium on Advanced Processing & Manufacturing Technologies (APMT) for Structural & Multifunctional Materials and Systems
Symposium 9	Porous Ceramics: Novel Developments and Applications

Symposium 10	Thermal Management Materials and Technologies
Symposium 11	Advanced Sensor Technology, Developments and Applications
Symposium 12	Materials for Extreme Environments: Ultrahigh Temperature Ceramics (UHTCs) and Nanolaminated Ternary Carbides and Nitrides (MAX Phases)
Symposium 13	Advanced Ceramics and Composites for Nuclear and Fusion Applications
Symposium 14	Advanced Materials and Technologies for Rechargeable Batteries
Focused Session 1	Geopolymers and other Inorganic Polymers
Focused Session 2	Computational Design, Modeling, Simulation and Characterization of Ceramics and Composites
Special Session	Pacific Rim Engineering Ceramics Summit

The conference proceedings are published into 9 issues of the 2011 Ceramic Engineering & Science Proceedings (CESP); Volume 32, Issues 2-10, 2011 as outlined below:

- Mechanical Properties and Performance of Engineering Ceramics and Composites VI, CESP Volume 32, Issue 2 (includes papers from Symposium 1)
- Advanced Ceramic Coatings and Materials for Extreme Environments, Volume 32, Issue 3 (includes papers from Symposia 2 and 12)
- Advances in Solid Oxide Fuel Cells VI, CESP Volume 32, Issue 4 (includes papers from Symposium 3)
- Advances in Ceramic Armor VII, CESP Volume 32, Issue 5 (includes papers from Symposium 4)
- Advances in Bioceramics and Porous Ceramics IV, CESP Volume 32, Issue 6 (includes papers from Symposia 5 and 9)
- Nanostructured Materials and Nanotechnology V, CESP Volume 32, Issue 7 (includes papers from Symposium 7)
- Advanced Processing and Manufacturing Technologies for Structural and Multifunctional Materials V, CESP Volume 32, Issue 8 (includes papers from Symposium 8)
- Ceramic Materials for Energy Applications, CESP Volume 32, Issue 9 (includes papers from Symposia 6, 13, and 14)
- Developments in Strategic Materials and Computational Design II, CESP Volume 32, Issue 10 (includes papers from Symposium 10 and 11 and from Focused Sessions 1, and 2)

The organization of the Daytona Beach meeting and the publication of these proceedings were possible thanks to the professional staff of ACerS and the tireless dedication of many ECD members. We would especially like to express our sincere thanks to the symposia organizers, session chairs, presenters and conference attendees, for their efforts and enthusiastic participation in the vibrant and cutting-edge conference.

ACerS and the ECD invite you to attend the 36th International Conference on Advanced Ceramics and Composites (http://www.ceramics.org/daytona2012) January 22-27, 2012 in Daytona Beach, Florida.

SUJANTO WIDJAJA AND DILEEP SINGH
Volume Editors
June 2011

Nanomaterials for Photocatalysis, Solar, Hydrogen, and Thermoelectrics

MORPHOLOGY CONTROLLED ELECTROSPINNING OF V_2O_5 NANOFIBERS AND THEIR GAS SENSING BEHAVIOR

R. von Hagen, A. Lepcha, M. Hoffmann, M. Di Biase and S. Mathur*

University of Cologne, Department of Chemistry
Institute of Inorganic Chemistry
Greinstrasse 6
D-50939 Cologne
Germany

ABSTRACT

Electrospinning of vanadyl acetylacetonate/polyvinylpyrrolidone hybrid nanofibers and calcination procedures design led to V_2O_5 nanofibers with control over their morphology. The fibers either consisted of densely joint nanorods or anisotropic nanocrystals forming porous nanofibers. They have been analyzed by thermogravimetric analysis, Fourier transform infrared spectroscopy, X-ray diffraction and scanning electron microscopy. Their dispersions in water/ethylene glycol could be used for inkjet printing of V_2O_5 gas sensors and the particle and film morphology influence on gas sensitivity was investigated.

Keywords: Electrospinning, V_2O_5, Nanorods, Gas-sensing

*e-mail: sanjay.mathur@uni-koeln.de, Tel: +49 (0)221 470 4107, Fax: +49 (0)221 470 4899

INTRODUCTION

Electrospinning is a versatile method for the production of 1D nanofibers. Originally designed for the production of simple polymeric fibers out of polymer solutions [1 a-d], the process was reinvented in the late 1990s for the production of functional nanofibers [2 a-b] by addition of specific compounds (metal salts, nanoparticles, drugs) into the polymeric precursor solution. The fiber materials range from biocompatible fibers for tissue-design to ceramic fibers for various applications such as energy storage and gas sensing.[2 b] The method allows an easy control over fiber morphology and alignment [3 a-b] and offers great perspective for fabrication of 1D nanomaterials and possible device integration. V_2O_5 is an intrinsic n-type wide optical band gap semiconductor with an electrical conductivity of around $0.5 \, S \, cm^{-1}$ at room temperature. The electrical conductivity is increased when part of the V^{5+} species is reduced to V^{4+} - accompanied by the simultaneous formation of oxygen vacancies - as the electron transport takes place in a hopping mechanism between V^{4+} impurities and V^{5+} centers.[4 a-b] Due to the unique redox behavior, the ability to form a wide spectra of homogeneously mixed-valent phases and the resulting electronic properties, V_2O_5 is of great interest for micro-, optoelectronic and gas-sensing devices. Several published reports suggest a good sensitivity and selectivity of V_2O_5 against reducing gases such as amines.[5 a-d] Furthermore its good chemical and thermal stability, the ability of multiple valency and the high redox-potential against Li/Li^+ makes it a promising cathode

3

material for Li-ion batteries with high capacity.[6] However, only a few examples of V_2O_5 nanofiber production by electrospinning can be found in literature. Differences in precursors, polymers or calcination procedures reportedly have a large influence on the V_2O_5 nanofiber morphologies, meaning that the method is able to produce defined nanocrystal morphologies like in wet chemical methods (e.g., solvothermal) with the advantage of no need of any workup steps.[7 a-c]

In this study we present the synthesis of V_2O_5 nanofibers by electrospinning using the low-cost vanadyl acetylacetonate ($VO(acac)_2$) precursor. By the design of calcination procedures we were able to design the V_2O_5 crystal morphology. The obtained nanofibers either consisted of dense joint nanorods or porously joint nanocrystals. The gas sensing properties of V_2O_5 nanoparticle films was comparatively analyzed by testing the sensitivity of inkjet printed gas sensors towards ethanol.

EXPERIMENTAL

The V_2O_5 precursor solution consisted of 0.375 mM ml^{-1} $VO(acac)_2$ and 37.5 mg ml^{-1} polyvinylpyrrolidone (PVP, 1.300.000 g mol^{-1}) in a mixture of dichloromethane/pyridine (8:2). After 3h of vigorously stirring the dark greenish homogeneous solution was transferred into a syringe and electrospun on a custom-built system consisting of a syringe pump (kdScientific, USA), motorized z-positioning (isel Germany AG, Germany), a high voltage source (HCN 35 – 35000 POS, F.u.G. Elektronik GmbH, Germany) and a grounded metallic collector. The syringe was connected to a needle (0.8 mm inner diameter) via a PTFE tube and set to a potential of 20 kV, the feeding rate was 20 µl min^{-1} and the needle-collector distance was 8 cm. After the spinning process the fibers were dried in vacuum over night and the following calcination procedure was done in two different ways. In the following the two different samples are named V_2O_5-1 and V_2O_5-2, respectively. V_2O_5-1 was calcined at 600 °C in air for 5 h and V_2O_5-2 was calcined for 5 h under Nitrogen at 500 °C and then postcalcined for 5 h at 600 °C in air. X-ray powder diffraction (XRD) patterns of the samples were measured on a STOE STADI MP diffractometer with Cu-Kα radiation ($\lambda = 1.5418$ Å). Fourier transform infrared (FT-IR) measurements were carried out on a Perkin Elmer Spectrum 400. The samples morphologies were observed with a FEI Nova NanoSEM 430 scanning electron microscope (SEM). Thermo gravimetric analysis was carried out on a METTLER Toledo TGA/DSC 1 STAR^e System. Ink-jet printing was done using a Microdrop MD-P-802 printer equipped with a nozzle of 50 µm inner diameter and a heatable stage which was set to 120 °C. The ink consisted of 4 mg V_2O_5-1 and V_2O_5-2, respectively, dispersed in a mixture of 4 ml H_2O/ethylene glycol (50:50 vol%), which could be printed at a voltage of 119 V and a pulse-length of 42 µs. The gas sensing property of the printed V_2O_5 nanocrystals was determined using a self designed measurement system operated at 220 °C using a Keithley 2400 source meter for resistivity measurement and controlled by LabView software.

RESULTS AND DISCUSSION

A thermo gravimetric analysis (TGA) of as-spinned $VO(acac)_2$/PVP composite nanofibers was done in air to define the calcination temperatures for the V_2O_5 formation. The heating rate was set to 10 °C min^{-1} and the starting weight 4.3 mg. The TGA (Fig. 1) showed two steps of large mass losses. The first weight loss (180 – 220 °C) step could be attributed to the thermal decomposition of organic ligands from $VO(acac)_2$ and condensation reactions. The second step (400 – 450 °C) possibly indicated the combustion/oxidation of residual organics and incipient crystallization of V_2O_5. The total weight loss was found to be 70 wt%.

a)
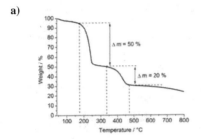

b)

Figure 1: a) Thermo gravimetric analysis of the as-prepared $VO(acac)_2$/PVP composite fibers carried out in air at a rate of 10 °C min^{-1}. b) SEM micrograph of the as-prepared fibers.

The SEM micrograph of the as-prepared nanofibers (Fig. 1 b) showed homogeneous fibers with a diameter of around 150 nm and bead formation at some spots, which is related to the viscosity of the solution. It has been shown that the rheological properties of the spinning solution are strongly influencing the fibers morphology and higher polymer concentrations by overall constant spinning parameters can suppress bead formation.[2b] The control over the V_2O_5 nanofiber morphology could be gained by two distinct calcination processes. The pyrolysis of green fibers for 5 h at 600 °C in air led to the formation of nanofibers consisting of fused V_2O_5 nanorods (V_2O_5-1) with an average diameter of 250 nm and an average length of 600 nm (Fig. 2).

a)   b)
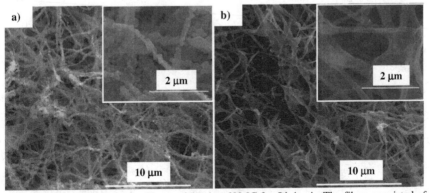

Figure 2: a) SEM image of the fibers calcined at 600 °C for 5 h in air. The fibers consisted of interconnected V_2O_5 nanorods. b) SEM image of the V_2O_5 nanofibers after calcination for 5 h at 500 °C in nitrogen followed by post calcination step at 600 °C for 5 h in air.

The formation of nanorods seemed to be thermodynamically favored as no further addition of structure-directing additives was needed. As similar observation after the calcination of NH_4VO_3/PVA fibers was reported by Mai et al.[7 c] They led the obtained V_2O_5 nanorod-in-nanowire formation back to the low solubility of their used NH_4VO_3 precursor in water, but as our precursor was well soluble in the chosen solvent mixture and the obtained results are similar the nanorod formation might be due to thermodynamic reasons in orthorhombic V_2O_5 crystallisation. The facile formation of V_2O_5 nanorods by electrospinning of low-cost V_2O_5 precursors and post-calcination treatment shows the potential of electrospinning, when compared to classical sol-gel or hydrothermal methods for V_2O_5 nanorod formation with tedious procedures involving several steps of synthesis and post-synthesis treatments are generally required.[5 a-c] Calcination of the as-prepared fibers under nitrogen at 500 °C for 5 h resulted in carbonized PVP/amorphous V_2O_5 composite fibers as shown by powder X-Ray diffraction (XRD) analysis (Fig. 3 a). After a second calcination step at 600 °C for 5 h in air, porous V_2O_5 nanofibers were obtained (V_2O_5-2) where the porous network was formed by intertwined V_2O_5 nanocrystals with an average size of 95 nm in length and 65 nm in diameter. The carbon out-burn of the previously formed composite fibers apparently led to simultaneous crystallization of the V_2O_5. Interestingly the nanocrystals formed in this procedure only showed a slight anisotropy, suggesting that the carbon incorporation or encapsulation present after calcination at 500 °C suppressed an elongated growth of the V_2O_5 crystals formed in the following second calcination step. The "manta ray" like structure of the as-prepared fibers was also visible in the V_2O_5-2 sample (Fig. 2 b) when the carbonized PVP, acting as a structural matrix, was removed.

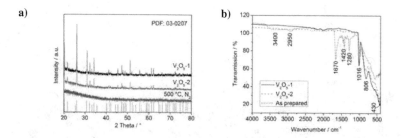

Figure 3: a) Powder XRD analysis of calcined fibers using different procedures. b) The FT-IR spectra of as-prepared and calcined V_2O_5 nanofibers.

The FT-IR spectra of the as-prepared fibers (Fig. 3 b) showed broad stretching bands for ν (OH) centered at 3400 cm⁻¹ and for ν (CH) at 2950 cm⁻¹, which can be attributed to PVP and absorbed water, respectively, the stretching bands at 1670 cm⁻¹, 1420 cm⁻¹ and 1280 cm⁻¹ can be attributed to ν (CO) and ν (CN) of the PVP and acetyl-acetonate ligands. The FT-IR of V_2O_5-1 and V_2O_5-2 showed the characteristic vibrational bands of V_2O_5 around 806-1016 cm⁻¹ which can be attributed to the stretching frequency of short V=O bonds. The broad vibrational band occurring at lower wavenumbers is attributed to the V-O-V octahedral bending modes.[8]

Dispersion of V_2O_5-1 and V_2O_5-2 fibers in a mixture of water/ethylene glycol (50:50 vol%) led to inkjet printable nanoinks which could be used for preparation of printed gas sensors. The sensors were fabricated following the procedure described in one of our recent publications [9] and tested towards the reducing gas ethanol to investigate the structural influence of differently V_2O_5 morphologies on the sensing performance. The different structural features of printed V_2O_5-1 and V_2O_5-2 are shown in Figure 4. The optical micrograph shows the printed compact linear structure made up by V_2O_5-nanoparticles which closed the electrical circuit of the interdigital Au-electrodes. The SEM images (Fig. 4 b and c) showed intrinsic differences in the grain size and porosity of the V_2O_5 nanoparticles in V_2O_5-1 and V_2O_5-2 which was mainly due to different calcination procedures. The film is more porous in the case of larger V_2O_5-nanorods (V_2O_5-1) than in the case of the smaller V_2O_5 nanoparticles (V_2O_5-2). It is known that the gas sensitivity of a material is strongly influenced by its size and structural dimensions, as the underlying electrical conductivity is a function of the particle morphology and grain sizes. Furthermore, it has been shown that the film structure (dense, porous) has an influence on the response and recovery time of the sensor.[10 a-b]

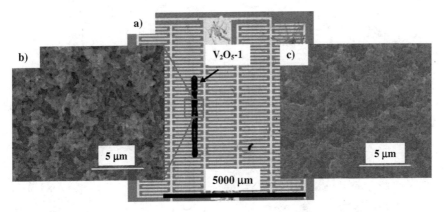

Figure 4: a) Optical micrograph of the printed gas sensor showing the general structure. b) SEM image of the printed V_2O_5-1 gas sensor showing the printed film consisted of V_2O_5-nanorods. c) SEM image of the printed V_2O_5-2 gas sensor showing the printed film consisted of almost isotropic V_2O_5-nanoparticles.

There are several reports available in the literature on the gas sensing behavior and mechanism of V_2O_5 and modified V_2O_5 nanostructures towards reducing gases especially amines.[5 a-c] In our experiment, the objective was to investigate the structural influence of the particle morphology/dimensions and film structure on the sensing performance. Therefore the sensitivity of V_2O_5-1 and V_2O_5-2 towards 500 ppm ethanol at 220 °C was recorded (Fig. 5). Both sensors showed a similar behavior as direct response towards the reducing gas ethanol and relatively long recovery time but in the case of more dense film made of smaller V_2O_5 nanocrystals (V_2O_5-2) an increase in sensitivity of 50 % ($\Delta(R_A/R_G) : \Delta(R_A/R_G)$) was observed compared to the larger nanorods based sensor (V_2O_5-1). As ethanol is reducing part of the V^{5+}

species and thereby increasing the number of V^{4+} atoms accompanying the formation of oxygen vacancies, the conductivity of the structures increased by introduction of ethanol into the sensing chamber.[5 a] As the V_2O_5 nanocrystals had smaller dimensions in the case of V_2O_5-2 the effect was stronger due to the larger surface to volume ratio and the stronger influence of the formed depletion layer on the conductivity at smaller grain sizes. Taking into account that the sensitivity was 50 % increased in case of V_2O_5-2 but the recovery time was the same as in V_2O_5-1, the more densely packed film of V_2O_5-2 showed a faster movement of charge carriers in the more entangled nanostructure.

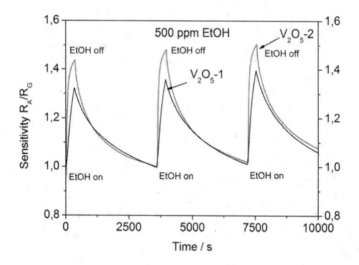

Figure 5: Sensitivity of printed V_2O_5 nanostructure based sensor towards 500 ppm ethanol at a working temperature of 220 °C.

CONCLUSION

The calcination of composite nanofibers fabricated from $VO(acac)_2$/PVP solutions led to different V_2O_5 nanofiber structures which composed of differently linked v_2o_5 nanocrystals which were differently shaped. We were able to produce nanofibers of densely joint V_2O_5 nanorods as well as porous V_2O_5 nanofibers. The V_2O_5 nanocrystals formed during clacination tent to show anisotropic growth and thereby forming nanorods, as they could be obtained after direct calcination in air. The two step calcinations under 1st nitrogen and 2nd air, led to the formation of porous V_2O_5 nanofibers build up from nanocrystals showing almost no anisotropy. The resulting structures are highly interesting for gas sensing as well as cathode materials for Li-ion batteries. Dispersion of the V_2O_5 nanofibers into water/ethylene glycol led to inkjet printable nanoinks which could be used to prepare printed gas sensors showing a direct response towards ethanol with a performance dependency regarding V_2O_5 nanoparticles as well as film morphology.

ACKNOWLEDGEMENT

The authors are thankful to the University of Cologne and the Federal Ministry of Education and Research (BMBF; KoLiWIn 55102006) for financial support.

REFERENCES

[1]   (a) J. F. Cooley, U.S. Patent 692.631. (b) W. J. Morton, U.S. Patent 0.705.691. (c) A. Formhals, U.S. Patent 1.975.504. (d) A. Formhals, U.S. Patent 2.349.950.

[2]   S. Ramakrishna, K. Fujihara, W. E. Teo, T. Yong, Z. Ma, R. Ramaseshan, *Materials Today*, **2006**, *9*, 40. (b) D. Li, Y. Xia, *Adv. Mater.*, **2004**, *16*, 1151.

[3]   (a) D. Li, Y. Wang, Y. Xia, *Nano Lett.*, **2003**, *3*, 1167. (b) D. Li, G. Ouyang, J. T. McCann, Y. Xia, *Nano Lett.*, **2005**, *5*, 913.

[4]   (a) O. Schilling, K. Colbow, *Sens. Actuators B.*, **1994**, *21*, 151. (b) J. Muster, G. T. Kim, V. Krstic, J. G. Park, Y. W. Park, S. Roth, M. Burghard, *Adv. Mater.*, **2000**, *12*, 420.

[5]   (a) K. I. Shimizu, I. Chinzei, H. Nishiyama, S. Kakimoto, S. Sugaya, W. Matsutani, A. Satsuma, *Sens. Actuators B*, **2009**, *141*, 410. (b) A. D. Raj, T. Pazhanivel, P. S. Kumar, D. Mangalaraj, D. Nataraj, N. ponpandian, *Curr. Appl. Phys.*, **2010**, *10*, 531. (c) I. Raible, M. Burghard, U. Schlecht, A. Yasuda, T. Vossmeyer, *Sens. Actuators B*, **2005**, *106*, 730. (d) S. Mathur, S. Barth, *Z. Phys. Chem.*, **2008**, *222*, 307.

[6]   N. A. Chernova, M. Roppolo, A. C. Dillon, M. S. Whittingham, *J. Mater. Chem.*, **2009**, *19*, 2526.

[7]   (a) P. Viswanathamurthi, N. Bhattarai, H. Y. Kim, D. R. Lee, *Scr. Mater.*, **2003**, *49*, 577. (b) C. Ban, N. A. Chernova, M. S. Whittingham, *Electrochem. Commun.*, **2009**, *11*, 522. (c) L. Mai, L. Xu, C. Han, X. Xu, Y. Luo, S. Zhao, Y. Zhao, *Nano Lett.*, **2010**, *10*, 4750.

[8]   J.C. Valmalette, J.R. Gavarri, *Mater. Sci. Eng. B*, **1998**, *54*, 168–173.

[9]   L. Xiao, H. Shen, R. von Hagen, J. Pan, L. Belkoura, S. Mathur, *Chem. Commun.*, **2010**, *46*, 6509.

[10]   (a) K. I. Choi, H. R. Kim, J. H. Lee, *Sens. Actuators B*, **2009**, *138*, 497. (b) F. Song, H. Su, J. Han, D. Zhang, Z. Chen, *Nanotechnology*, **2009**, *20*, 495502.

FABRICATION OF NANOSTRUCTURED α-Fe$_2$O$_3$ FILMS FOR SOLAR-DRIVEN HYDROGEN GENERATION USING HYBRID HEATING

Bala Vaidhyanathan[*,¶], Sina Saremi-Yarahmadi[¶], K.G. Upul Wijayantha[#]
[¶]Department of Materials, Loughborough University, Leicestershire, LE11 3TU, UK

ABSTRACT

Electrodeposited thin films of Fe were oxidised using a novel conventional/microwave hybrid heating method. The photo-performance of hematite electrodes was investigated and the results are compared with regards to the amount of microwave power applied. The findings showed significant improvement in the performance of hematite electrodes when microwave heating was used. The genuine 'microwave effect' observed in this case is confirmed by using hybrid heating experiments at identical time-temperature profiles. The photocurrent density obtained at 0.23 V vs. $V_{Ag/AgCl}$ increased significantly from 7 to 126 μA.cm^{-2} when microwave power was raised from 0 to 300 W. The films prepared by pure conventional annealing showed high recombination and photocurrent onset of around 0.4 V vs $V_{Ag/AgCl}$ while the other showed a negative shift to 0.1 V vs $V_{Ag/AgCl}$ for the hybrid samples. The results obtained from Raman spectroscopy indicated a highly defective crystalline nature for the conventionally-annealed samples while microwave-assisted annealing resulted in fewer defects in the oxygen sublattice of hematite structure. It suggests that microwave heating improves surface properties of hematite films thus enhancing the photoelectrochemical performance of the photoelectrodes. Hybrid heating was found to provide a unique opportunity to control/tailor the oxidation kinetics and in turn the photo-performance of hematite electrodes using microwave power.

INTRODUCTION

Solar hydrogen generation through water splitting is one of the most promising renewable environmentally-friendly energy technologies. Since the earliest reports of photoelectrochemical (PEC) water splitting by Fojishima and Honda in 1971 [1], significant efforts have been made in order to develop efficient water splitting photoelectrodes by adopting different approaches such as investigating various material systems (e.g. metal oxides [2-6], sulphides [7-8]) as well as fabrication procedures (e.g. CVD [9,10], electrodeposition [11,12] and sputtering [13,14]).

Hematite (α-Fe$_2$O$_3$) is a strong candidate photoelectrode for PEC water splitting as it meets most of the selection criteria of a suitable photocatalyst material for this application such as bandgap, chemical and PEC stability, and ease of fabrication. However, one of the major barriers in the development of efficient hematite photoelectrodes is the short hole diffusion length in hematite (2-4 nm) hence poor hole transport at the hematite/electrolyte interface[15,16]. As a result, photocurrent densities and conversion efficiencies reported for undoped hematite electrodes have not been promising. In order to overcome this and improve the photo-performance, different approaches such as doping [6,9,11,13] and/or formation of nanostructures with sizes comparable to the hole diffusion length have been implemented, some of which have led to significant improvements in photo-performance of hematite electrodes [6,9].

The processing of materials using microwaves has seen significant interest in the last few years. The majority of research is conducted using domestic or modified microwave ovens and a wide spectrum of materials have been synthesised most notably carbides, nitrides, complex oxides, phosphates and silicides [17]. Recently, we reported a facile route for the fabrication of iron oxide photoelectrodes by utilising microwave energy [18]. We showed for the first time that the conversion of iron into iron oxide using a microwave assisted process occurred at lower temperatures and at faster times than what had been previously reported [18]. It was also demonstrated that the less demanding processing conditions associated with the microwave approach combined with the specific advantage

11

of rapid heating and cooling was beneficial for the fabrication of α-Fe$_2$O$_3$ thin films as a result of retained nanostructure and minimised grain coalescence [18]. However, the specific role played by microwaves in enhancing the photo-performance was not investigated in detail. Here, we examine the effect of microwave power on the performance of nanostructured hematite photoelectrodes with the aim of elucidating a correlation between the microwave power and the performance. This is probed by applying varying amounts of microwave energy during the oxidation of Fe films while keeping every other experimental parameter exactly the same. The results are compared with the performance of the film fabricated solely through a conventional heating method. The performance of the films prepared under microwave-assisted conditions exceeded that of the pure conventional photoelectrode confirming the genuine effect of microwave energy on PEC performance enhancement.

EXPERIMENTAL

Fe films were deposited on fluorine-doped tin oxide (FTO) substrates (TEC8, Pilkington Glass Ltd., St Helens, UK) using an electrochemical route previously reported [18] where Tetraethoxysilane TEOS) was used as the source of Si. The area of the deposited film was maintained at 1 cm^2. The thermal oxidation of Fe films were carried out in three different methods. Firstly, the Fe films were annealed using purely conventional and purely microwave methods. The details of the conventional furnace and microwave oven together with the heating procedure employed are presented elsewhere [18]. The third heating method which was designed to investigate the effect of microwave power was conducted in a conventional/microwave hybrid furnace (Fig. 1(a)). The description of the furnace is given elsewhere [19]. The furnace was capable of operating in pure conventional or microwave/conventional hybrid mode; the microwave frequency used was 2.45 GHz, and up to 2 kW of microwave power was available if required. The microwave power level was fixed at different values of 0, 100, 200 and 300 W and the conventional power used varied to provide the same sintering schedule for each of the experiments. The films were heated to 500 °C and soaked at this temperature for 30 minutes. Typical temperature-time profiles employed are given in Fig. 1(b), all samples are subjected to identical time-temperature profiles, with microwave power being the only variable. The furnace was then turned off and the films were left to cool. Throughout the work, temperature was controlled using a Luxtron optical fibre thermometer (M10, Luxtron Corporation, Santa Clara, CA).

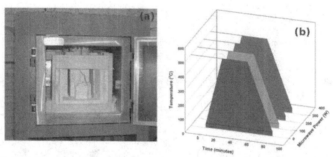

Fig. 1. (a) Hybrid microwave furnace. (b) Typical temperature-time plots employed for the oxidation of Fe films.

PEC performance of hematite electrodes was measured using a three-electrode configuration in 1 M NaOH electrolyte, Ag/AgCl/KCl (3M) as reference electrode, and a platinum wire as a counter electrode. The potential of the photoelectrode was controlled by a potentiostat (microAutoLab, type III, Windsor Scientific, Berkshire, UK). The light source was an AM 1.5 class A solar simulator provided by Solar Light (Solar Light 16S – 300 solar simulator, Solar Light Company Inc., PA, USA). The distance between the light and the electrode was maintained in such a way that the intensity of the light incident on electrode was at 1000 W.m^{-2}. The intensity of the light was measured using a PMA2144 Solar Light pyronometre (Solar Light Company Inc., PA, USA). The incident photon to electron conversion efficiency (IPCE) was obtained by measuring the incident photon flux using a 75 W Xenon lamp connected to a monochromator (TMc300, Bentham Instruments Ltd., Berkshire, UK). The light was calibrated using a silicon photodiode. Photocurrent spectra were measured at 0.5 V vs. V$_{Ag/AgCl}$ using a combination of a lock-in amplifier (Bentham 485, Bentham Instruments Ltd., Berkshire, UK) and an in-house built potentiostat. Readings were collected at 5 nm intervals while the monochromated light was scanned from 320 to 650 nm. Raman spectroscopy was performed using a HORIBA Jobin Yvon LabRAM HR (with 632.8 nm He-Ne laser) Raman spectrophotometer (HORIBA Jobin Yvon Ltd., Middlesex, UK). The spectra were obtained in the range of 100 to 900 cm^{-1}. Surface nanostructure of the films were studied using Leo 1530 VP (Carl Zeiss NTS Ltd., Cambridge, UK) field emission gun SEM (FEGSEM) at an accelerating voltage of 5 kV.

RESULTS AND DISCUSSION

The PEC performance of the hematite photoelectrodes was studied by photocurrent density vs. potential (J-V) characteristics. Initially, two types of samples were characterised: samples annealed under conventional heating and sample annealed using pure microwave heating. The J-V of the best performing samples (conventional heating at 450 °C for 15 minutes and microwave heating at 250 and 270 °C for 15 minutes in the case of undoped and Si-doped samples, respectively) are compared in Fig.2. It is evident that the samples annealed using microwave heating showed significant improvements in PEC performance. In the case of undoped samples, the photocurrent density achieved at 0.23 V vs. V$_{Ag/AgCl}$ raised from ~60 to ~110 μA.cm^{-2} when microwave energy was used as the source of heating. Similar trend was observed for Si-doped samples however the difference in the photocurrent density was slightly smaller, from 145 to 180 μA.cm^{-2}.

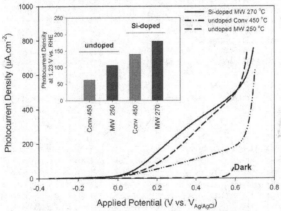

Fig. 2. Photocurrent density – potential plots of undoped and Si-doped hematite electrodes prepared using either conventional and microwave heating methods. Inset compares the photocurrent density of different photoelectrodes at 1.23 V vs. RHE (0.23 V vs. V$_{Ag/AgCl}$).

The results of the photoperformance exhibit two major advantages when the microwave heating was used to oxidize the Fe films. Firstly, microwave heating resulted in a considerable reduction in the temperature required to obtain enhanced performance. The annealing temperature was reduced from 450 to 270 °C. The other interesting aspect of microwave heating was the sizable enhancements in the photocurrent density obtained from the hematite electrodes. The increase in the performance was consistent in both undoped and doped samples where a maximum two-fold increase in the photocurrent density at 0.23 V vs. $V_{Ag/AgCl}$ was observed. At higher potentials, where charge transport and collection progressed more efficiently, the enhancement in the photoperformance was even higher.

Surface morphology of the samples prepared with conventional and microwave heating was studied using FEG-SEM. The SEM micrographs obtained for pure conventional and pure microwave undoped hematite films are shown in Fig. 3. Comparing the two micrographs, the different in the surface morphology and nanostructure of the two types of photoelectrodes are evident. The sample treated with microwave heating (Fig. 3(b)) exhibits a surface morphology comprising of much smaller particles when compared to the conventionally-annealed sample (Fig. 3(a)). The conventionally-annealed sample was formed from larger particles which were densly sintered together resulting in the formation of well-connected agglomerated particulates with sizes exceeding 200 nm. This had significant implications to the photo-performance of hematite electrodes. The finer the nanostructure and surface morphology of the hematite films the better the performance. Other factors such as crystallinity and surface properties of the photoelectrode also play major roles in altering the performance of hematite films.

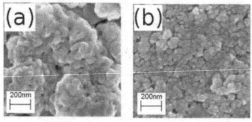

Fig.3. SEM micrograph of undoped hematite films using (a) conventional heating at 450 °C for 15 minutes and (b) microwave heating at 250 °C for 15 minutes.

In order to elucidate the exact role of microwave heating and the reasons behind the enhancements of photo-performance a series of experiments were conducted in which the only changing parameter was the level of microwave power used during the heating procedure. By controlled changes to the microwave power level used, it was possible to establish a correlation between microwave power and the extent of oxidation of the hematite films. The effect of microwave energy on the PEC performance of hematite films was studied using the J-V curves of photoelectrodes. As shown in Fig.4, the sample which was fabricated without the influence of microwave heating (pure conventional sample) had the poorest PEC performance among all the samples. While the samples prepared using hybrid heating showed photocurrent densities beyond 100 μA.cm^{-2} at 0.3 V vs. $V_{Ag/AgCl}$ (e.g. 183 μA.cm^{-2} for 300 W microwave), the pure conventional sample showed only 16 μA.cm^{-2} at the same potential. Fig.5 illustrates the correlation between increasing microwave power and the photocurrent density obtained at 0.23 V vs. $V_{Ag/AgCl}$, which showed a linear increase in the PEC performance as the microwave power is raised. The highest photocurrent density achieved in this potential was about 126 μA.cm^{-2} compared to a negligible 7 μA.cm^{-2} observed for pure conventional sample.

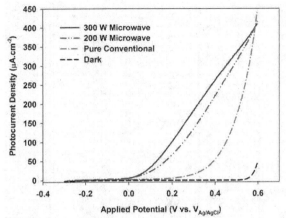

Fig. 4. Photocurrent density – potential plots of hematite electrodes prepared using different levels of microwave energy. The photocurrent density – potential curve under dark conditions is shown by a dashed line.

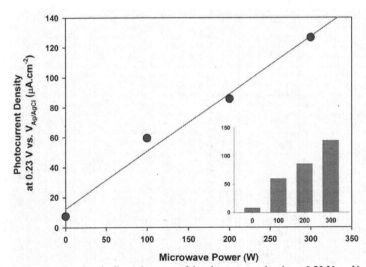

Fig. 5. The plot showing the linear increase of the photocurrent density at 0.23 V vs. $V_{Ag/AgCl}$ of hematite photoelectrodes as a function of microwave power applied. The inset shows a bar chart comparing the photocurrent density at the same potential.

The onset of the photocurrent density, where the steep increase in the photocurrent starts, was dramatically reduced as the amount of microwave power present increased. The photocurrent onset is observed at around 0.4 V vs. $V_{Ag/AgCl}$ for pure conventional sample while it reduced to ~0.1 and 0.03 V vs. $V_{Ag/AgCl}$ for 200 and 300 W microwave treated samples. The occurrence of the more positive onset potential in case of pure conventional sample can be regarded as an indication of large number of recombination centres at the surface of hematite electrodes and the accumulation of electronic and ionic surface charge at the hematite/electrolyte interface [20]. This positive displacement of the onset potential as a result of slow hole transfer kinetics would lead to a loss of free energy and hence reduced efficiency [20]. Microwave-assisted heating appears to improve the surface properties of the hematite electrodes and enhance the PEC performance of hematite photoelectrodes.

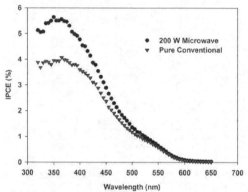

Fig. 6. Quantum efficiency (IPCE) spectra of two samples prepared without and with microwave energy, pure conventional and 200 W, respectively. The measurement potential is 0.5 V vs. $V_{Ag/AgCl}$.

The quantum efficiency of different films was studied using IPCE measurements. It provides useful information on the maximum photon to electron conversion efficiency that can be obtained for each sample at different wavelengths. Fig.6 depicts the IPCE spectra of two samples, namely pure conventional and 200 W hybrid-heated samples. The measurement potential was selected at 0.5 V vs. $V_{Ag/AgCl}$ in order to obtain a better spectrum for the pure conventional sample. As shown, there is no clear difference between the IPCE of both types of samples at wavelengths higher than 570 nm which is beyond the absorption threshold of hematite; the significant changes occur at lower wavelengths. The maximum efficiency observed (at 350 nm) was 5.6 % and 3.9 % for the 200 W microwave and pure conventional sample, respectively.

The light penetration depth at short wavelengths is small and very close to the top surface of the films. As explained earlier, one of the major drawbacks of hematite films is the short hole diffusion length. The holes that are generated at the vicinity of the surface of the electrode will be collected more efficiently at the surface as they have to travel shorter distances to reach the hematite/electrolyte interface. However, if the recombination centres at the surface restrict the charge transfer the efficiency will be reduced. The differences observed in the IPCE of the samples is in agreement with the photocurrent density characteristics observed in Fig.4, suggesting a larger degree of surface recombination in the pure conventional sample compared to that of the samples prepared under the presence of microwave energy.

Raman spectroscopy measurements were carried out in order to obtain a clearer insight into the surface properties of the photoelectrodes. The Raman spectra for different samples are shown in Fig. 7. Raman spectrum of a single crystal of hematite exhibits seven peaks at about 226 (A$_{1g}$), 245 (E$_g$), 293 (E$_g$), 298 (E$_g$), 412 (E$_g$), 500(A$_{1g}$) and 612 (E$_g$) cm^{-1}, which are due to the transverse optical modes (TO modes) mentioned in parentheses [21]. All photoelectrodes prepared in this study show hematite peaks at ~226, 245, 298, and 400 cm^{-1}. In the case of the sample annealed using pure conventional heating the peak at 500 cm^{-1} is very weak and the Raman band at 612 cm^{-1} seems to be superimposed by very strong peak at ~670 cm^{-1}. The latter had been observed in our previous work also [18] and was attributed to the either incomplete oxidation and/or the presence of lattice disorder. As the microwave power level increases the peak at 612 cm^{-1} becomes more defined and the relative intensity of the 670 cm^{-1} peak decreases.

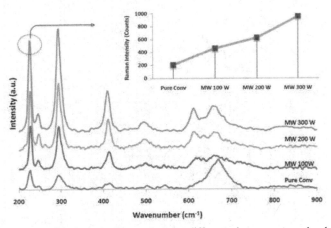

Fig. 7. Raman spectra of the hematite films prepared at different microwave power levels. Inset plot shows the intensity of the Raman peak at 229 cm^{-1} in different samples.

It has been reported previously that as the particle size decreases, the Raman peaks become broader and shift towards lower wavenumbers [22]. As shown in Fig.6, the broadest peaks are observed for the pure conventional sample and as the microwave power increases the intensity of the Raman peaks becomes stronger. However, there is no considerable difference between the full-width-half-maximum (FWHM) of any of the main hematite peaks in the three samples annealed using microwaves. However, the shifts in the Raman bands were not consistent. Basically, peaks at low wavenumbers, namely 226, 245 and 298 cm^{-1}, do not shift at all and the wavenumber at which the maximum intensity occurs remains unchanged in all the samples. However, the other peaks at 412, 500 and 612 cm^{-1} shift towards lower wavenumbers as the microwave power decreases. Due to this inconsistency, we conclude that the difference in the particle size of films in this study is minimal.

The peaks at lower wavenumbers are associated with the movements of the Fe-ions while the other three Raman bands at higher wavenumbers are related to the movements of the oxygen ions [21]. These results indicate that oxygen sublattice (extent of oxidation) is influenced by the presence of the microwave power. Therefore it can be concluded that the oxygen sublattice were affected by the microwave power and the absence of the microwave power resulted in the formation of a hematite structure with more oxygen deficiency. In other words, pure conventional heating did not influence the highly defective nature of the oxygen sublattice while harnessing microwave energy resulted in the less structural defects. It seems that the contributing factor to high surface recombination in the pure conventional sample is the highly defective oxygen sublattice as it provides ideal recombination sites. The formation of oxygen deficient sublattice in isostructural α-Cr_2O_3 has been confirmed by Raman spectroscopy before [23], however, the origin of them was due to the size dependant structural changes. However, Cherynshova et al showed that in the case of hematite, the oxygen sublattice is relatively resistant to size dependant structural changes and contributed to the strong restructuring of the metal sublattice [22]. The results obtained here suggest that microwave heating could influence the oxygen sublattice. This resulted in a less defective oxide structure and improved the crystallinity of the samples both of which contributed to a better PEC performance for these photoelectrodes. The high microwave absorption characteristics of Fe particles ensures rapid and uniform heating which could lead to bulk nucleation/oxidation. Effective utilisation of reaction enthalpy and the magnetic field of the microwave radiation may also be playing a role in enhancing the oxidation kinetics.

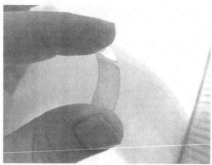

Fig. 8. Iron oxide films deposited on flexible transparent substrates.

These results suggest that microwave heating can improve surface properties of hematite photoelectrodes thus improving the performance. The fact that these changes can be implemented under isothermal conditions and by changing the microwave power is extremely useful as this provides a new means of tailoring the performance of photoelectrodes. Along with the reductions in the annealing temperatures required for obtaining efficient photoelectrodes, this has the potential to open up new possibilities for the fabrication of photoactive films on plastic, flexible substrates. Currently, work is underway to fabricate hematite films on flexible substrates (Fig.8) and develop efficient photoelectrodes by utilising microwave heating methodology. In addition, further characterisation is needed to fully understand the structural changes of hematite as a result of microwave heating. It is also important to investigate the properties of the interface between the hematite film and the FTO substrate as a function of microwave energy used for processing.

CONCLUSIONS

Photoelectrodes of hematite are fabricated by electrodeposition of Fe films and subsequent oxidation to iron oxide using conventional and hybrid microwave/conventional methods. By varying the level of the microwave power used in hybrid heating, the effect of microwave energy on improving the performance of hematite photoelectrodes is examined. It is shown that the photocurrent density at 0.23 V vs. $V_{Ag/AgCl}$ increased significantly from 7 to 126 $\mu A.cm^{-2}$ when microwave power was varied from 0 to 300 W. The poor PEC performance of the photoelectrodes prepared by pure conventional heating is correlated to the highly defective crystal structure and associated charge recombination. The films prepared by pure conventional annealing exhibited photocurrent onset of around 0.4 V vs. $V_{Ag/AgCl}$ while the onset shifted to lower potentials of less than 0.1 V vs. $V_{Ag/AgCl}$. Investigation of the surface properties of the photoelectrodes using Raman spectroscopy revealed that samples annealed under microwave irradiation consist of less defective oxygen sublattice in the hematite structure which seems to contribute to the better PEC performance. Further investigation is required to elucidate the exact mechanisms involved during the microwave heating and the corresponding effects on the properties of the hematite/substrate interface.

ACKNOWLEDGMENTS

S. Saremi-Yarahmadi thanks Materials Research School at Loughborough University for providing the PhD studentship. Authors are also grateful for the help of Prof J. Binner, Prof R. C. Thomson, Dr K Yendall, Dr K Annapoorani and ceramic group researchers at Department of Materials, Loughborough University. We are also thankful to Mr B Dennis at Department of Physics and Dr A. Tahir at Department of Chemistry, Loughborough University for helpful discussions.

* Corresponding Author, Bala Vaidhyanathan, E-mail: B.Vaidhyanathan@lboro.ac.uk,
Tel: +44(0)1509 223152. Fax: +44(0)1509 223949.
Department of Chemistry, Loughborough University, LE11 3TU, UK

REFERENCES
[1] A. Fujishima., and K. Honda, Eelectrochemical Photolysis of Water at a Semiconductor Electrode, *Nature,* **238**, 37-38 (1972).
[2] P. Kamat, Meeting the Clean Energy Demand: Nanostructure Architectures for Solar Energy Conversion, *J. Phys. Chem. C*, **111**, 2834-2860 (2007).
[3] H. Wang, T. Lindgren, J. He, A. Hagfeldt, and S.-E. Lindquist, Photoelectrochemistry of Nanostructured WO₃ Thin Film Electrodes for Water Oxidation: Mechanism of Electron Transport, *J. Phys. Chem. B*, **104**, 5686-5696 (2000).
[4] B. Yang, P.R.F. Barnes, W. Bertrama, and V. Luca, Strong Photoresponse of Nanostructured Tungsten Trioxide Films Prepared via Sol-gel Route, *J. Mater. Chem.*, **17**, 2722-2729 (2007).
[5] V.R. Satsangi, S. Kumari, A.P. Singh, R. Shrivastav, and S. Dass, Nanostructured Hematite for Photoelectrochemical Generation of Hydrogen, *Int. J. Hydrogen Energy*, **33**, 312-318 (2008).
[6] R. van de Krol, Y. Liang, and J. Schoonman, Solar Hydrogen Production with Nanostructured Metal Oxides, *J. Mater. Chem.*, **18**, 2311-2320 (2008).
[7] A. Deshpande, and N.M. Gupta, Critical Role of Particle Size and Interfacial Properties in The Visible Light Induced Splitting of Water Over The Nanocrystallites of Supported Cadmium Sulphide, *Int. J. Hydrogen Energy*, **35**, 3287-3296 (2010).
[8] J.F. Reber, and M. Rusek, Photochemical Hydrogen Production with Platinized Suspensions of Cadmium-Sulfide and Cadmium Zinc-Sulfide, *J. Phys. Chem.*, **90**, 824-834 (1986).
[9] I. Cesar, A. Kay, J. A. Gonzalez Martinez, and M. Gratzel, Translucent Thin Film Fe₂O₃ Photoanodes for Efficient Water Splitting by Sunlight: Nanostructure-Directing Effect of Si-Doping, *J. Am. Chem. Soc.*, **128**, 4582-4583 (2006).
[10] A.A. Tahir, K.G.U. Wijayantha, S. Saremi-Yarahmadi, M. Mazhar, and V. McKee, Nanostructured α-Fe₂O₃ Thin Films for Photoelectrochemical Hydrogen Generation, *Chem. Mater.*, **21** 3763-3772 (2009).

[11] A. Kleiman-Shwarsctein, M.N. Huda, A. Walsh, Y. Yan, G.D. Stucky, Y.–S. Hu, M.M. Al-Jaseem, and E.W. McFarland, Electrodeposited Aluminum-Doped α-Fe₂O₃ Photoelectrodes: Experiment and Theory, *Chem. Mater.*, **22**, 510-517 (2010).

[12] H.E. Prakasam, O.K. Varghese, M. Paulose, G.K. Mor, and C.A. Grimes, Synthesis and Photoelectrochemical Properties of Nanoporous Iron (III) Oxide by Potentiostatic Anodization, *Nanotechnology*, **17**, 4285 (2006).

[13] J.A. Glasscock, P.R.F. Barnes, I.C. Plumb, N. Savvides, Barnes, I.C. Plumb, and N. Savvides, Enhancement of Photoelectrochemical Hydrogen Production from Hematite Thin Flms by the Introduction of Ti and Si, *J. Phys. Chem. C*, **111**, 16477–16488 (2007).

[14] P. Hiralal, S Saremi-Yarahmadi, B.C. Bayer, H. Wang, S. Hofmann, U. Wijayantha, and G.A.J. Amaratunga, Nanostructured Hematite Photoelectrochemical Electrodes Prepared by The Low Temperature Thermal Oxidation of Iron, *Sol. Energy Mater. Sol. Cells*, In Press, (2011)

[15] C.J. Sartoretti, B.D. Alexander, R. Solarska, W. A. Rutkowska, J. Augustynski, and R. Cerny, Photoelectrochemical Oxidation of Water at Transparent Ferric Oxide Film Electrodes, *J. Phys. Chem. B*, **109** 13685-13692 (2005).

[16] J.H. Kennedy, and K.W. Frese, Photo-Oxidation of Water at Alpha-Fe₂O₃ Electrodes, *J. Electrochem. Soc.*, **125**, 709-714 (1978).

[17] K.J. Rao, B. Vaidhyanathan, M. Ganguli, and P.A. Ramakrishnan, Synthesis of Inorganic Solids Using Microwaves, *Chem. Mater.*, **11**, 882-895 (1999).

[18] S. Saremi-Yarahmadi, B. Vaidhyanathan, and K.G.U. Wijayantha, Microwave-Assisted Low Temperature Fabrication of Nanostructured α-Fe₂O₃ Electrodes for Solar-Driven Hydrogen Generation, *Int. J. Hydrogen Energy*, **35**, 10155-10165 (2010).

[19] J. Wang, J. Binner, B. Vaidhyanathan, N. Joomun, J. Kilner, G. Dimitrakis, and T.E. Cross, Evidence for the Microwave Effect During Hybrid Sintering, *J. Am. Ceram. Soc.* **89**, 1977-1984 (2006).

[20] K.G.U. Wijayantha, S. Saremi-Yarahmadi, and L.M. Peter, Kinetics of Oxygen Evolution at a-Fe₂O₃ Photoanodes: A Study by Photoelectrochemical Impedance Spectroscopy, *Phys. Chem. Chem. Phys.*, In Press, (2011) DOI: 10.1039/C0CP02408B.

[21] I.V. Chernyshova, M.F. Hochella Jr, and A.S. Madden, Size-Dependent Structural Transformations of Hematite Nanoparticles. 1. Phase Transition, *Phys. Chem. Chem. Phys.*, **9**, 1736-1750 (2007).

[22] W.B. White, The Structure of Particles And The Structure of Crystals: Information From Vibrational Spectroscopy, *J. Ceram. Process Res.*, 6, 1-9 (2005).

[23] J. Zuo, C. Xu, B. Hou, C. Wang, Y. Xie, and Y. Qian, Raman Spectra of Nanophase α-Cr₂O₃, *J. Raman Spectrosc.*, 27, 921-923 (1996).

Nanostructured Membranes, Thin Films, and Functional Coatings

SYNTHESIS AND CHARACTERIZATION OF BIMETAL DECORATED CARBON SPHERES FOR SENSING APPLICATIONS

Innocentia V Sibiya* and S. Sinha Ray

DST/CSIR Nanotech. Innovation Centre, National Centre for Nano-Structured Materials, Council for Scientific and Industrial Research, Pretoria 0001, Republic of South Africa.
*Corresponding author. E-mail: ISibiya@csir.co.za

ABSTRACT

Carbon nano-spheres (CNSs) were synthesised using the direct pyrolysis of a hydrocarbon gas at high temperatures. The CNSs were then functionalized and decorated with bimetallic particles of gold (Au) and silver (Ag) using microwave assisted catalyst synthetic procedures. The as synthesised, functionalised and Au/Ag decorated CNSs were extensively characterised by scanning electron microscopy (SEM), X-ray diffraction (XRD), thermogravimetric analysis (TGA) and Raman spectroscopy. SEM studies revealed how the sizes of the particles vary with varying concentrations of hydrocarbon gas. TGA analysis revealed the information on the thermal stability of CNSs and it composites with Au/Ag. XRD analysis exhibited the formation of cubic structures of the bimetals on the CNSs surfaces. Raman spectra showed the degree of graphitization was improved at higher concentrations of the bimetals on spheres surfaces.

INTRODUCTION

Gas detection and monitoring are important aspects since they create an immediate and long term health risk to personnel involved, whether it is in industrial, medical or commercial applications. Common gas sensors are metal oxide semiconductors such as tin oxides, zinc oxide, titanium oxide and aluminium oxide. Problems encountered with these sensors are: lack of flexibility, poor response time and they are operated at elevated temperatures. Therefore, a new material is needed for the fabrication of small, user-friendly and reliable gas sensing device [1].

Over the last few years, extensive research has been carried out to design small but cheap gas sensors which possess high sensitivity, selectivity and stability. This is where nanotechnology comes to the picture. Nanotechnology is an advanced technology dealing with materials having a range from 1nm to 100nm. The development of nanotechnology has created a huge potential to build highly sensitive, low cost, portable sensors with lower power consumption. The extremely high surface–to-volume ratio and hollow structure of some nanomaterials is ideal for the adsorption of gas molecules [2].

Carbon nanotubes (CNT's) are cylindrical carbon molecules that have a unique geometry, morphology, and properties. They posses very unique characteristics due to their hollow centre, nanometer size and large surface area, and are able to change their electrical resistance drastically when exposed to alkalis, halogens and other gases at room temperature. Hence, they have a potential to be better chemical sensors [3].

CNSs produced by chemical vapour deposiotion are advantageous for some applications where high porosity, high surface area, and /or a high degree of graphitization are desired. So these CNSs can be substituted for CNT's, which are typically expensive to manufacture [4].

The CNSs have useful properties such as unique shape, size and or electrical properties. The absence of functional groups on the surface of the CNSs is believed to be responsible for some of the beneficial and novel properties of the CNSs. For example, carbon nanomaterials having reduced functional groups have shown improved dispersibility in many organic polymers [5].

EXPERIMENTAL

CNSs.Synthesis Using Chemical Vapour Deposition (CVD)

In this experiment, CNSs were produced by CVD method without a catalyst. The temperature of the furnace was set to 800°C under nitrogen flow rate of 240 ml/min. As the temperature reached 800°C, then acetylene (C_2H_2) with a flow rate of 200 ml/min was flushed as the carbon source for 5 min and then switched off. This reaction occurred in a quartz tube inserted in the furnace.

Microwave Synthesis

In a typical procedure, 1g of as-synthesized CNSs was oxidized with 5M HNO_3 in a microwave reaction system (Anton- Paar Multiwave 3000) for 5 min at 120 C. The solution was filtered, washed and the precipitate was dried at 110 C overnight. In the second step, various amounts of (1.5, 2.5, 3.5, 4.5 and 5.5 wt %) Au/Ag metals were loaded onto 0.2g O- CNSs surfaces. The loading of the metals onto CNSs was done in a microwave for 5 min at 60°C. This step was optimized in order to get uniform distribution of metals on the CNSs surfaces (see Table 1). The precipitate was then removed by centrifugation, washed several times with distilled water and dried at 110°C overnight.

Table1. Average diameter of spheres at different reaction time at constant Temperature and carbon source flow rate

Average diameter /nm	Carbon source reaction time /min
150-250	20
254-362	15
198-227	10
97.5-141	5

RESULTS AND DISCUSSION

Fig. 1 shows the SEM images of as synthesised, oxidised and various Au/Ag decorated CNSs. The size dependancy of the spheres on the reaction time is shown in Table.1,and we assume that longer the carbon source flow reaction time, the rougher the surface and the shorter the carbon source flow reaction time, the smoother the surface [6]. A rough surface is more favourable for attachments than smooth surfaces because there are more docking sites for metal particle attachment than with smooth surfaces. Spheres that were synthesised at 5min have a smooth surface with a diameter of (97.5-141) nm; the spheres were oxidised before being functionalised. Spheres obtained were not well separated and the reason for this is that pyrolysis gives rise to chemical combination causing agglomeration of the carbon spheres [7]

In this case surface roughness and smoothness will be confirmed by AFM for further studies. Based on SEM images the as-synthesised CNS show a slightly rough surface while the oxidised CNS show a smooth surface, due to all the amorphous carbon being removed during oxidation. The SEM image of CNS loaded with 1.5 wt% of bimetalllic particles (BMP's) show square, triangular and spherical BMP's that don't look well dispersed on the surface of the carbon spheres (CS). The same patterns of poor dispersion of BMP's were observed with little improvement; when the BMP's weight percentage was increased from 1.5 wt% to 2.5 wt% BMP on the CNS surface.

The size of the BMP's formed were of a micrometer and not of a nanometer scale which means that the 1.5wt% ratio of BMP's selected was high and a weight percentage below 1% should be considered or tried to see if the production of bimetallic nanoparticles can be possible by lowering the concentration, since optimum temperature and flow rate of carbon source that favour ripening were obtained.

As the ratio of BMP's on CNS increase the dispersion, shapes and sizes of BMP's differ from image to image. Observations showed that on CNS loaded with 3.5wt% of BMP's square BMP's were formed, while with 4.5wt% of BMP's triangular prisms were formed, and with 5.5wt% of BMP's the shape is not clear.

The morphology of these crystals depends on the distance of the formation conditions from thermodynamic equilibrium. Meng et al. showed that increasing the driving force for crystallisation in this case being the concentration of BMP's results in the crystals shape varying the way they do [8].

TGA traces of weight loss as a function of as-synthesised CNS, O-CNS, O-CNS 1.5BMP, O-CNS2.5BMP, O-CNS3.5BMP, O-CNS4.5BMP and O-CNS5.5BMP were measured both under nitrogen and air atmospheres. TGA results are presented in Fig. 2. Results show that CNSs are more stable under both environments.

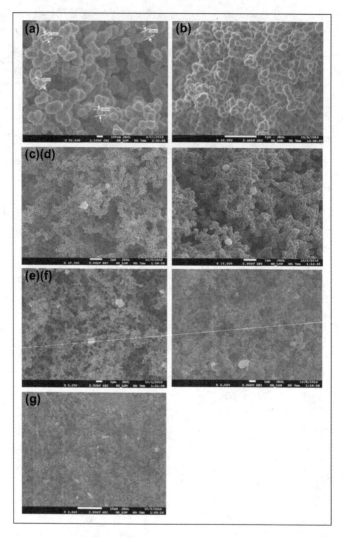

Figure 1. SEM images of (a) as synthesised CNSs (b) oxidised CNSs (O-CNSs) (c) O-CNS -1.5BMP
(d) O-CNS-2.5BMP (e) O-CNS-3.5BMP (f) O-CNS-4.5BMP (g) O-CNS-5.5BMP

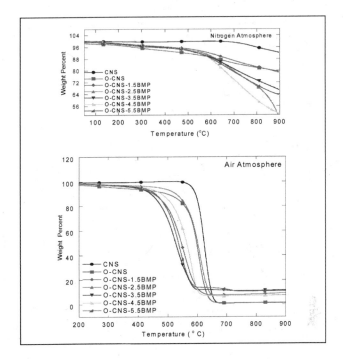

Figure 2: TGA scans of various samples under nitrogen and air atmospheres

Under nitrogen atmosphere two steps of degradation was observed for the composites: at 588 °C was the first degradation, which was due to the carbon loss. A little bit of char and Au was found on the pans after the run, this was because sample degradation is very slow under nitrogen atmosphere.

Under air atmosphere the as-synthesised degrades at 563°C, oxidised at 516°C and the composites degrade faster and at lower temperatures than they do under nitrogen atmosphere. This is because the silver metal under air atmospheres at high temperatures corrodes; making the oxidation process to occur faster than it does under nitrogen atmosphere, degrading the carbon and the silver and leaving out the gold. No char was observed but remains of Au were observed, and this is because gold is non-reactive but will only increase thermal conductivity.

Overall the CNS with high loadings of bimetals show very high mass loss and low onset temperatures in both atmospheres compared to the as synthesised CNS and the ones loaded with low composition of bimetals on them, this is also due to the high percentage of silver that promotes oxidation and makes the degradation rate faster.

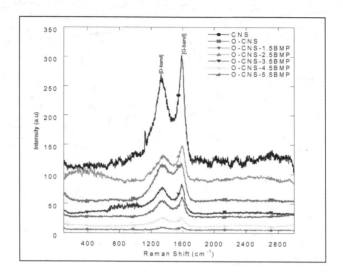

Figure. 3 Raman spectra of as synthesised CNSs, oxidised CNSs, and various Ag/Au loaded CNSs.

Fig. 3 shows the Raman spectra of as synthesised CNSs, O-CNSs, and various Ag/Au loaded CNSs. The I_D and the I_G average ratios derived from peak areas were 0.85, 0.85, 0.83, 0.84, 0.68, 0.83 and 0.69 which corresponds with CNS, O-CNS, O-CNS1.5BMP, O-CNS2.5BMP, O-CNS3.5BMP, O-CNS4.5BMP and O-CNS5.5BMP respectively. The appearance of the bands is consistent with the presence of defects and disorder in the samples [9]. Raman spectra revealed that the intensity ratio of the bands, i.e., I_D / I_G decreases with increase in concentration of the BMP ratio loaded. This also suggests that the degree of graphitisation improves at higher BMP loadings. Overall since all the values are below one, it shows that CNS are highly graphitic with only a few defects.

Fig. 4 shows the strong diffraction peaks at the Braggs angles of 38°,43°,65° and 77° correspond to the (111), (200) (220) and (311) planes, of face centered cubic of Ag and Au crystals, that occur on the same positions [8]. The presence of peaks other than the ones at 25° and 82° which are because of the graphite layers of the CNS, are due to the synthesised silver chloride (AgCl) which was formed and confirmed by the XRD database used to identify peaks. These results also showed that the Ag and Au BMP's were immobilised in a dispersed way on the surface of the CNS. A better reducing agent and stabilizer for both silver nitrate and gold chloride hydrate using microwave assisted synthesis which will stabilize both silver nitrate to Ag and gold chloride hydrate to Au without forming AgCl that is suggested is sodium citrate [10]. Using this reducing agent will atleast ensure that all the chlorides and nitrates are removed during centrifugation since all the Ag and Au formed will be in a stable form.

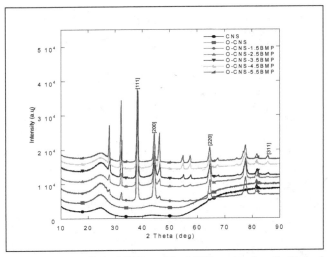

Figure 4. XRD patterns of as synthesised CNSs, oxidised CNSs, and various Ag/Au loaded CNSs.

CONCLUSIONS

Carbon spheres of smooth surface from 97.5-141 nm were successfully produced using the CVD method without a catalyst support. The CNS was decorated with various amounts of Ag/Au using microwave assisted synthesis. SEM images showed that after oxidation the surface became smooth. TGA analysis revealed that CNS show a higher thermal stability under both nitrogen and air environments due to their high stability at temperatures up to 600°C, while composites showed less due to the ability of Ag to oxidise making the oxidation process faster. In Raman the degree of graphitisation of particles and the plain CNS does not differ that much and since all of them gave values less than one, it can be said that they are highly graphitic. XRD analysis showed that Ag and Au are present but due to the method used during synthesis, some silver chloride peaks act as a surface poison, blocking the active sites. Using a better reagent that will stabilize the metals without forming AgCl may enable the development of CS-based gas sensors; if weight percentages of bimetals loaded on the CNS were lower that one.

REFERENCES

1. Koh, S.K Jung, H.J.,Song, S.K., Choi, W.K., Choi, D., Jeon, J.S (2000). Sensor having tin oxide thin film for detecting methane gas and propane gas, and process for manufacturing theof, US patent 6,059,937
2. Md Yasin Faizah, *European Journal of Scientific Research*, 35 (2009) 142-149.
3. Yun Wang and John T.W. Yeow, *Journal of Sensors* (2009) doi: 10.1155/2009/493904
4. Cheng Zhang., Martin Fransson., Bing Zhou., (2010). Method for manufacturing carbon nanostructures having minimal surface functional groups, US patent 7,718,156
5. Cheng Zhang., Martin Fransson.,Changkun Liu., Bing Zhou., (2010). Carbon nanostructured from catalytic templating nanoparticles, US patent 7,718,155
6. Shaochun Tang , Yuefang Tang, Sascha Vongerhr, Xiaoning Zhao, Xiang Kang Meng, *Applied Surface Science* 225 (2009) 6011-6016
7. Qilang Lin, Minzhi Zheng, Tao Qin, Rongrong Guo, Penghui Tian, *Journal of Analytical and Applied Pyrolysis.* 89 (2010) 112-116
8. X.K Meng, S.C. Tang and S. Vongehr , *Journal of Material Sciences and Technology,* 26 (2010) 487-522
9. S.D. Mhlanga , N.J Coville, S.E Iyuke, A.S Afolabi, A.S Abdulkareen, N. Kunjuzwa, *Journal of Experimental Nanoscience,* 5 (2010) 40-51
10. Rongjing Cui., Chang liu., Jianming Shen., Di gao., Jun-Jie Zhu and Hong-Yuan Chen., *Advanced Functional Materials* 18 (2008) 2197-2204

Nanotubes and Polymer Nanocomposite Technology

MICROWAVE IRRADIATION OF RUTHENIUM ON NITROGEN-DOPED CARBON NANOTUBES

Letlhogonolo F. Mabena[a,b*], Suprakas Sinha Ray[a], Neil J. Coville[b]

[a]DST/CSIR Nanotech. Innovation Centre, National Centre for Nano-Structured Materials, Council for Scientific and Industrial Research, Pretoria 0001, South Africa.
[b]School of Chemistry, University of the Witwatersrand, Johannesburg, South Africa.
*Corresponding author. E-mail: TMabena@csir.co.za;

ABSTRACT

Results obtained on microwave irradiation method to prepare Ruthenium (Ru) nanoparticles on nitrogen doped carbon nanotubes (N-CNTs) are reported. The N-CNTs were prepared by chemical vapour deposition. X-ray diffraction, high resolution transmission electron microscopy (HR-TEM) and Raman techniques were used to characterise the Ru/CNT material. The results show that the microwave assisted method allows synthesis to be achieved using shorter reaction times, reduced energy consumption than conventional heating. The results of HR-TEM show a good dispersion of metals on the N-CNTs. Ru nanoparticles with very narrow size distribution and small particle size with an average diameter of 2.5 nm on N-CNTs were obtained.

INTRODUCTION

Catalysis plays an important role in the preparation of compounds in numerous fields such as fuels and fine chemicals [1, 2]. Advances in catalysis field have shown that nanoparticles with nanostructured surfaces enhance the catalytic activity as compared to larger particles [3, 4]. The challenges for catalyst improvement are the design of the structural factors such as morphology, nanoparticle size and inter-particle distance. [5].

Intensive research has been conducted in order to develop new catalyst supports which can modify the catalytic activity and the selectivity of the existing catalytically active phase. The interaction of nanoparticles with a support plays an important role in the design of the catalyst.[2]. Carbon based nanostructures (e.g. carbon nanotubes, carbon spheres, carbon fibers, etc.) have been found to be good supports for metal nanoparticles [2, 6]. Studies have shown that doping carbon nanostructures with nitrogen generate n-type semi conductors, which have greater electron mobility than the corresponding undoped carbon nanostructures [7-9]. Nitrogen introduces chemically active sites that are required for anchoring of nanoparticles deposition [10]. Therefore, it is proposed that the improved catalytic activity of nitrogen containing carbon nanostructures is due to the higher dispersion of metals and good interaction between the support and the metal particles [11]. In addition, nitrogen is able to generate defects on carbon, which increases the edge plane exposure and as a result enhance the catalytic activity [12].

The dispersion of a metal on a support is still largely based on conventional catalyst preparation techniques, such as wet impregnation followed by chemical reduction. These approaches often lack fine control of structural factors. In addition, use of these processes can be time consuming since they include multiple steps such as long ageing, drying and calcination of the samples. Hence, there is an interest in using alternative techniques to synthesise catalyst such as by microwave irradiation [13, 14]. In this paper, we report on the deposition on ruthenium (Ru) nanoparticles on the nitrogen doped carbon nanotubes (N-CNTs) using a microwave polyol assisted deposition method.

EXPERIMENTAL

Reagents and Pre-Treatment of Pristine CNTs

$RuCl_3.xH_2O$, ethylene glycol (EG), HNO_3 and pristine CNTs with inner diameter of 5-10 nm and outer diameter of 40-80 nm were obtained from Sigma Aldrich. Pristine CNTs were functionalised with 30 % HNO_3 at 110 C for 2 h and dried in an oven at 110°C for overnight.

N-CNTs synthesis with two stage furnace by thermal-chemical vapour deposition (CVD)

N-CNTs were synthesised using a thermal-CVD method in a horizontal split-tube furnace. The reactions were carried out in a tubular quartz reactor (28 mm in diameter). Cyclohexanol was used as carbon source, aniline as a nitrogen source and ferrocene as catalyst. A Mixture of cyclohexanol-aniline-ferrocene in a 1:5:34 ratio was placed in a quartz boat that was directly introduced in the centre of the first furnace and vaporised at 280°C. The resultant vapours were transferred to the second furnace where the N-CNTs were grown at a temperature of 900°C under the nitrogen carrier gas flow. The reaction was maintained until there was no longer vapour visible in the tube; the furnace was cooled down under the carrier gas atmosphere. The black soot was then collected from the inner wall of the quartz tube and treated with 30 % HNO_3 at 110°C for 2 h and was dried in an oven at 110°C overnight.

Deposition of Ru nanoparticles on CNTs

Ru nanoparticles were deposited on CNTs (pristine CNTs and N-CNTs) using a microwave assisted polyol and conventional reflux polyol methods.

In a microwave teflon vessels, 50 mg of CNTs, 10 mg $RuCl_3.xH_2O$ were mixed with 80 ml EG and sonicated for 10 min to afford homogenous suspension. The suspension was the then placed in a microwave reactor (Anton-Paar Multiwave 3000). The suspension was heated to 200 C and kept at the temperature for 5 min at various powers settings: 350, 500, 800 and 1000 W.

Same, $RuCl_3.xH_2O$/CNTs/EG method described above was was mixed in a conical flask and refluxed in an oil bath for 24 h. at 200 C.

After reactions the resulting suspensions from both microwave and conventional heated reaction were filtered and the residue was washed with 50 ml acetone, then with 200 ml deionised water and dried at 110 C overnight.

CHARACTERISATION

The transmission electron microscopy (TEM) analysis was performed by using a JEOL2100 microscope with beam energy of 200 kV. TEM was used to analyse the particle size and morphology of the as-prepared CNTs. The samples for analysis were prepared by sonicating about 0.5 mg of CNTs in 5 ml methanol for 5 min. One drop of the resulting suspension was dropped on a copper grid. Raman spectra were collected before and after synthesis using a Horiba Jobin Yvon T64000 Raman. The spectra was recorded from 150 - 3000 cm^{-1} with a 514.5 nm excitation with energy setting 1.2 mW from a Coherent Innova model 308 Ar ion laser in measured backscattering geometry with a cooled charge-coupled device array detector.

The morphology of the CNTs was evaluated using a LEO 1525 FE-SEM. The SEM was equipped with energy dispersive spectroscopy (EDS) facility, which was used for elemental analysis of the CNTs. Thermal analyses were performed by simultaneous thermogravimetric-differential using a TA Q500 thermal instrument. CNTs samples (5 mg) were analysed in platinum pans at a heating rate of 10 °C/min to 950 °C in an atmosphere of air flowing at 50 ml/min.

RESULTS AND DISCUSSION

Fig.1a shows the SEM images of pristine CNTs. There is external diameter ranged between 7-15 nm. N-CNTs with outer diameter of 50-90 nm and length of 13 μm are shown in Fig 1b. The TEM image of N-CNTs, fig 1b (inset) exhibits the typical 'bamboo like' compartments which are promoted by the presence of nitrogen [15, 16].

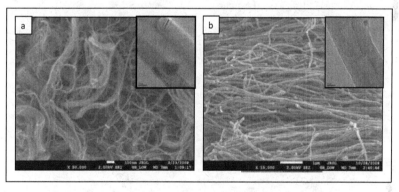

Figure 1. SEM images of the pristine (a) CNTs and the (b) N-CNTs

Raman spectroscopy was used to measure the graphitisation and the degree of defects. Material shows the disorder peak around 1350 cm^{-1} corresponding to a D-band and a G-vibration modes peak at 1580 cm^{-1}. The D-band is attributed to the disorder-induces feature due to the lattice distortion, G-band originate from the in plane stretching vibration mode E_{2g} of a single crystal graphite, [17]. The peak at 2702 cm^{-1} represents second order G' band, an overtone mode of the D band which correspond to two - phonon processes and its appearance does not need disorder [18].

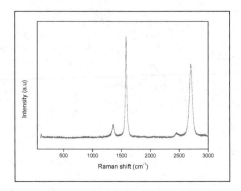

Figure 2. Raman spectrum of the pristine CNTs before and after treatment with acid.

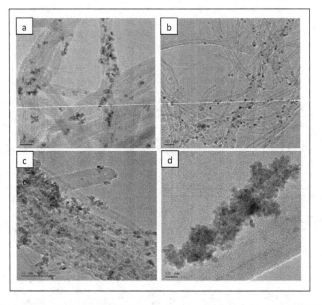

Figure 3. TEM images of Ru/CNTs prepared at different microwave power settings. (a) 350 W; (b) 500 W; (c) 800 W; (d) 1000 W

The TEM analysis of Ru/CNTs prepared using different microwave power settings (refer Fig. 3) and with sodium citrate stabilizer (Fig. 4). In order to study the effect of microwave power, a series of samples were prepared by heating to 200°C at 350, 500, 800 and 1000 W respectively. At a power setting of 350W the particles were not homogenously dispersed on the surface of the CNTs. A power setting of 500W was found to be optimum as it gave ultra-fine metallic Ru nanoparticles homogenously dispersed on the surface of the CNTs with diameter ranging from 5 – 8 nm in Fig. 2(b). As the power was increased above 500 W the particles formed agglomerates with uneven distribution. The agglomerates formed at higher power settings made it difficult to determine the size of the particles. Since power is directly proportional to time the agglomeration of the particles is attributed to the rapid heating rate of the reaction, which does not provide sufficient time for the reduction of the metal ions.

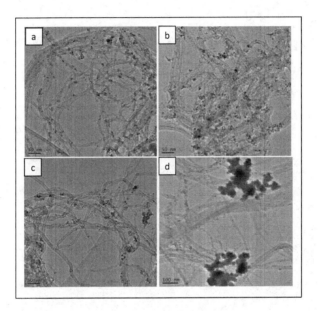

Figure 4. TEM images Ru/CNTs with sodium citrate stabilizer: (a) 0%; (b) 10 %; (c) 30%; (d) 50%. (Microwave setting 500W)

The effect of citrate stabilizer was also investigated. Fig. 4 shows that with 0 and 10% of sodium citrate stabilizer the nanoparticles hardly aggregate and they are well dispersed on the surface of CNTs, while with the increase of the stabilizer Ru particles began to agglomerate and considerable amount of clusters are observed. Even though the particles were agglomerating with increase with citrate, few that were free showed decrease in size.

The XRD patterns of the Ru on pristine CNTs with and without citrate stabilizer are shown in Fig. 5. The diffraction peaks at $2\theta = 26.6°$, $54.2°$ and $78.1°$ are associated with the (002), (004) and (110) planes [19-21]. The other four peaks are characteristics of the hexagonal crystalline Ru [19, 21]. The FWHM of plane Ru (101) is widening with the addition of citrate which implies the decrease in the particle size of Ru. Guo et al [22] and Sakthivel et al [6] reported similar effects in the preparation of metal nanoparticles using citric acid and SB12 as stabilizers with microwave.

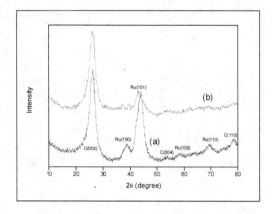

Fig. 5. XRD patterns for Ru deposited on pristine CNTs (a) with and (b) without citrate

The particle size distribution histogram of Ru/CNTs is shown in Fig 5. The histogram was done by counting 250 particles on TEM image. Without citrate stabilizer [Fig.5 (a)] the average diameter of the Ru particles on MWCNTs is 5 nm. When Stabilizer was used the size distribution moves to a lower nanometer range with a mean diameter of 4nm [Fig. 5(b)].

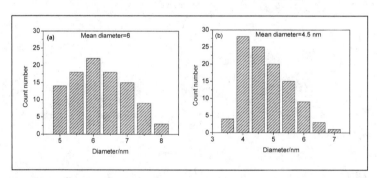

Figure 6. Particle size distribution histogram of Ru/CNTs prepared with 500 W: (a) No citrate; (b) with citrate

N-CNTs produced were subjected to 30% HNO_3 treatment to remove the Fe particles and other impurities. After acid treatment, a significant increase in the intensity of a D-band can be observed in Raman spectra in Fig 7. This suggests the two-dimensional graphitization is altered to a more disordered structure due to the damage of the wall when the metal impurities from synthesis are removed on the wall surface. This was proved by higher I_D/I_G ratio of acid-treated N-CNTs of 0.63 in comparison with 0.52 ratio of the untreated N-CNTs. Thermogravimetric analysis (TGA) results showed that certainly there was a removal of impurities (Fig. 7). There was less than 10 wt% residual left after treatment with acid from initial 20 wt%. The remaining might be the Fe_2O_3 metals particles inside the tubes which are not easy to remove.

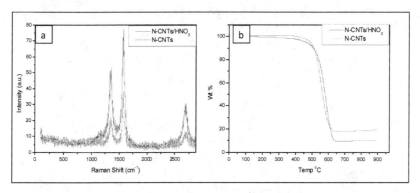

Figure 7. (a) Raman and (b) TGA profile of N-CNTs before and after treatment with acid.

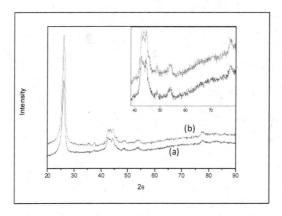

Figure 8. XRD patterns for (a) N-CNTs and (b) Ru/N-CNTs

In Fig. 8 the XRD pattern shows the diffraction lines of the carbon only and there is no difference with the Ru/N-CNTs diffraction pattern. The result implies that the sizes of the metallic Ru formed are below 4 nm detectable limitation of XRD. All the patterns were the same for both conventional and microwave heating.

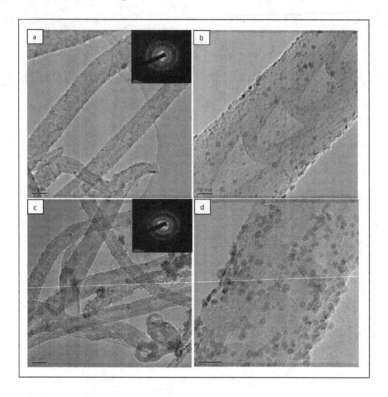

Figure 9. TEM images of Ru deposited on N-CNTs (a) – (b) Microwave irradiation (c) - (d) conventional heating.

Fig. 9 shows the TEM micrographs and the selected area electron diffraction (SAED) patterns (insets) of the Ru deposited on N-CNTs. SAED diffraction of both nanoparticles shows continuous rings corresponding to the CNTs collectively with some spots, which are due to metallic Ru. The EDS analysis shows that the particles are composed of Ru. It is clear from the TEM image that homogenously highly dispersed spherical particles were attached to the surface of the support with narrow size distribution ranging from 1 - 3 nm when microwave was used. Conventional heating had a good dispersion with slightly bigger distribution particle size of 3 - 4.5 nm and with minor metal aggregates. The average particle sizes derived from the TEM images are 2 nm for microwave irradiation and 3.5 nm for conventional heating. Microwave being a closed system build its process faster in comparison to conventional heating and it is accepted that the rate of reduction determine the size of the metal particle. These contribute to smaller particles without clusters. The high dielectric constant of EG also contributed to the rapid heating under microwave irradiation [14, 23, 24]. At the same time, well dispersion is due to the contribution of the nitrogen atoms on the surface of the nanotubes. Nitrogen atoms entering the graphene sheets as substitutes for carbon modify the adsorption strength of the nanotube towards foreign elements which in turn, will greatly modify the overall catalytic activity as well as selectivity [19, 25].

CONCLUSIONS

The microwave-assisted of RuCl₃ and CNTs in EG had evidently promoted the formation and homogenously dispersion of Ru nanoparticles with very narrow size distribution and small particle size with an average diameter of 2.5 nm for N-CNTs and 5 nm for pristine CNTs. Microwave irradiation is faster and more efficient than the conventional heating as the conventional heating produced particles slightly bigger with few clusters. The sizes of nanoparticles were optimized by varying, microwave heating power, conventional heating time and the ratio of the stabilizer to Ru precursor. In conclusion N-CNTs are the excellent support when coupled with the efficient method of deposition.

ACKNOWLEDGEMENTS

TM and SSR thank the DST and CSIR for financial support.

REFERENCES

[1] B. M. Reddy, K. N. Rao, G. K. Reddy, P. Bharali, Journal of Molecular Catalysis A: Chemical 253 (2006) 44-51.
[2] M.-J. Ledoux, C. Pham-Huu, Catalysis Today 102-103 (2005) 2-14.
[3] S. Martínez-Méndez, Y. Henríquez, O. Domínguez, L. D'Ornelas, H. Krentzien, Journal of Molecular Catalysis A: Chemical 252 (2006) 226-234.
[4] K. Patel, S. Kapoor, D. Dave, T. Mukherjee, Journal of Chemical Sciences 117 (2005) 53-60.
[5] G.-Y. Gao, D.-J. Guo, H.-L. Li, Journal of Power Sources 162 (2006) 1094-1098.
[6] M. Sakthivel, A. Schlange, U. Kunz, T. Turek, Journal of Power Sources 195 7083-7089.
[7] A. A. Koós, M. Dowling, K. Jurkschat, A. Crossley, N. Grobert, Carbon 47 (2009) 30-37.
[8] J. L. Figueiredo, M. F. R. Pereira, Catalysis Today 150 2-7.
[9] M. S. Saha, R. Li, X. Sun, S. Ye, Electrochemistry Communications 11 (2009) 438-441.
[10] R. I. Jafri, N. Rajalakshmi, S. Ramaprabhu, Journal of Power Sources 195 8080-8083.
[11] K. Chizari, I. Janowska, M. Houllé, I. Florea, O. Ersen, T. Romero, P. Bernhardt, M. J. Ledoux, C. Pham-Huu, Applied Catalysis A: General 380 72-80.
[12] T. C. Nagaiah, S. Kundu, M. Bron, M. Muhler, W. Schuhmann, Electrochemistry Communications 12 338-341.
[13] J.-Y. Kim, K.-H. Kim, S.-H. Park, K.-B. Kim, Electrochimica Acta 55 8056-8061.
[14] X. Li, W.-X. Chen, J. Zhao, W. Xing, Z.-D. Xu, Carbon 43 (2005) 2168-2174.
[15] K. Ghosh, M. Kumar, T. Maruyama, Y. Ando, Carbon 48 191-200.
[16] P. H. Matter, E. Wang, U. S. Ozkan, Journal of Catalysis 243 (2006) 395-403.
[17] E. Xu, J. Wei, K. Wang, Z. Li, X. Gui, Y. Jia, H. Zhu, D. Wu, Carbon 48 3097-3102.
[18] L. G. Bulusheva, A. V. Okotrub, I. A. Kinloch, I. P. Asanov, A. G. Kurenya, A. G. Kudashov, X. Chen, H. Song, physica status solidi (b) 245 (2008) 1971-1974.

[19] R. Chetty, S. Kundu, W. Xia, M. Bron, W. Schuhmann, V. Chirila, W. Brandl, T. Reinecke, M. Muhler, Electrochimica Acta 54 (2009) 4208-4215.
[20] B. Li, C. Wang, G. Yi, H. Lin, Y. Yuan, Catalysis Today In Press, Corrected Proof.
[21] D.-J. Guo, Journal of Power Sources 195 7234-7237.
[22] J. W. Guo, T. S. Zhao, J. Prabhuram, C. W. Wong, Electrochimica Acta 50 (2005) 1973-1983.
[23] D. M. Han, Z. P. Guo, R. Zeng, C. J. Kim, Y. Z. Meng, H. K. Liu, International Journal of Hydrogen Energy 34 (2009) 2426-2434.
[24] W.-X. Chen, J. Y. Lee, Z. Liu, Materials Letters 58 (2004) 3166-3169.
[25] J. Amadou, K. Chizari, M. Houllé, I. Janowska, O. Ersen, D. Bégin, C. Pham-Huu, Catalysis Today 138 (2008) 62-68.

AMINE FUNCTIONALIZATION OF CARBON NANOTUBES FOR THE PREPARATION OF CNT
BASED POLYLACTIDE COMPOSITES-A COMPARATIVE STUDY

Sreejarani K. Pillai,* James Ramontja, Suprakas Sinha Ray
DST/CSIR NIC, National Centre for Nano-Structured Materials
Council for Scientific and Industrial Research
Pretoria, Gauteng, South Africa

ABSTRACT
        This work describes a comparison between two chemical functionalization strategies for the
amine functionalization of carbon nanotubes (CNTs). In the first procedure, the CNTs are
functionalized in direct amination process that avoids the use of strong acids or acid chloride unlike the
conventional functionalization methods whereas the second method is a two step procedure involving
mild acid treatment followed by amidation. Both procedures allow not only to control of amine content
on the CNTs surfaces' but also to obtain remarkable degree of functionalization. The functionalization
of CNTs is confirmed by analytical techniques like scanning electron microscopy, Fourier transform
infrared spectroscopy, and X-ray photoelectron spectroscopy. The modified CNTs with optimum amine
content are used to prepare Polylactide (PLA) /CNT nanocomposites with improved properties through
solution casting method.

INTRODUCTION
        Carbon nanotubes (CNTs) and the novel carbon based nanomaterials have been the subject of
world wide research interest in recent years. They are proven to have unique electronic, mechanical,
and physical properties.[1,2] However, the limited solubility of CNTs in most organic solvents limits their
chemical manipulation, quantitative characterization, and wide applications. In recent years, there has
been much interest in preparing homogeneous dispersions/solutions of CNTs, suitable for processing
into thin films and composites exploiting the unrivalled properties of CNTs. The main routes consist of
end and/or sidewall functionalization, use of surfactants with sonication or high-shear mixing, [3-6]
polymer wrapping of nanotubes [7-10] and protonation by superacids.[11] Among all the methods, grafting
of CNT surface with amines has been widely investigated in preparing soluble CNTs. Wong et al.
reported modification of multi-walled CNTs (MWCNTs) via amide bond formation between carboxyl
functional groups bonded to the open ends of MWCNTs and amines.[12] Chen et al.[13] have demonstrated
that full-length single-walled CNTs (SWCNTs) can be solubilized in common organic solvents by
noncovalent (ionic) functionalization of the carboxylic acid groups by using octadecyl amine (ODA).
They found that the same dissolution process applied to arc-produced MWCNTs, average length < 1
   m), only gave rise to very unstable suspensions in organic solvents which were visually scattering. It
has been shown by Qin et al. [14] that by modifying Haddon's method using two Soxhlet extractors, large
quantities of solubilized MWCNTs could be prepared. However, the conventional approach of amine
functionalization is tedious with a typical reaction time of 4-8 days which involves steps such as
carboxylation, acyl chlorination followed by amidation.   Although these methods are quite successful,
they often indicate chopping of the tubes into smaller pieces (may be due to the oxidation induced
cutting during refluxing with concentrated acid and acid chloride for a long time) thus partly losing the
high aspect ratio (length/diameter) of CNTs. For the structural applications such as nanotube-based
composites, full-length MWCNTs are preferred because of their high aspect ratio. Hence incorporation
of nanotubes without losing the structural integrity and homogeneously dispersing them in a polymer
matrix still remains a great scientific challenge.
        The motivation for the current work is to develop simple strategies for the amine
functionalization of MWCNTs (referred as CNTs throughout the text) outer surfaces' that improves
their dispersion in a Polylactide (PLA) matrix. The work is focused on single and two step hexadecyl

amine (HDA) functionalization of CNTs where the amine content on CNT surface is successfully controlled by varying the reaction time. For a comparison, conventional amine functionalization which involves concentrated acid and thionyl chloride treatment was also performed. The effectiveness of functionalization procedures were analyzed using scanning electron microscopy (SEM), Fourier-transform infrared (FTIR) and X-ray photoelectron (XPS) spectroscopies. The degree of dispersion/distribution of CNTs in PLA was revealed by SEM and the thermal stability of the prepared composites was studied by thermogravimetric analyses (TGA).

EXPERIMENTAL

Amine functionalization of CNTs
    The CNTs used in this study were purchased from Sigma Aldrich with more than 95% purity (inner diameter-10 nm, outer diameter-20nm, length-0.5 to 500 m). Hexadecylamine (HDA), chloroform, ethanol and HNO_3 were purchased from Sigma-Aldrich and used as received.
In a single-step procedure, 0.2 g of as received CNTs was refluxed with 1 g of HDA in a round bottomed flask at an optimized temperature of 180 C in an oil bath. The degree of amine functionalization was controlled by varying the reaction time. The excess of HDA was removed by washing with ethanol several times. The solid was collected by Nylon membrane filtration (0.45 m pore size), and dried at 110°C overnight (referred to as CNTs-1).
    In a two-step functionalization procedure, as received CNTs were oxidized by refluxing with 5M HNO_3 (10 ml/0.1 g) for 1 h filtered, washed and dried at 110°C overnight (referred to as O-CNTs). The oxidized MWCNTs (0.2 g) were then refluxed with 2 g of HDA at an optimized temperature of 120 °C by varying the reaction time. The solution was filtered, washed with ethanol to remove excess HDA dried at 110°C overnight (referred to as CNTs-2).
    The weight percentage of HDA in the functionalized samples was calculated from gravimetric analysis using the equation: [15] Weight % of HDA = (Weight of HDA in the sample/Weight of the sample) 100. The functionalized CNTs thus obtained were sonicated in 100 ml chloroform for 30 min to obtain a black solution which was used for the PLA-CNT composite preparation.

Preparation of PLA/CNT composites
    PLA/CNT composites containing 1.5 wt% of functionalized CNTs (CNTs-1 and CNTs-2) were prepared by using a solution-blending film casting method. A known amount of PLA was first dissolved in a minimum amount of chloroform at 50 C. 100 ml of CNT solution in chloroform containing a predetermined amount of amine functionalized CNTs was sonicated with chloroform solution of PLA in an ultrasonic bath for 2h. The mixture was then cast on a glass petri dish and kept at room temperature for a day to evaporate the solvent which resulted in black composite film. The composite film was subsequently dried at 70 C under vacuum for 2 days. A neat PLA film, in the absence of CNTs, was also prepared by using the same technique. Dried neat PLA and PLA/CNT composite films were chopped into pieces and compression moulded by pressing under 2 MPa pressure at 180 C for 10 min. These compression moulded films were used for the various characterizations.

Sample Characterization
    The surface morphology of the solid samples and composite film was studied by a Leo 1525 FE-SEM using 6 kV accelerating voltage. All the powder samples were sputter coated with carbon to avoid charging. The nanocomposite sample was fractured in liquid nitrogen and then coated with carbon. FT-IR spectra were measured in transmittance mode by a Perkin Elmer Spectrum 100 FT-IR spectrometer (diamond crystal mode). XPS analyses were performed on a Kratos Axis Ultra device, with a monochromatic Al X-ray source (1486.6 eV). Survey spectra were acquired at 160 eV and

region spectra at 20 eV pass energies. Thermal property of the samples was investigated by TG analysis using a Q500 TGA instrument. The samples were heated in platinum crucibles under the air flow of 50 ml min^{-1}. The dynamic measurement was between ambient and 1000° C with a ramp rate of 10 ° C min^{-1}.

RESULTS AND DISCUSSION

The functionalized CNTs were black powders. The weight percentage of HDA in the CNTs-1 and CNTs-2 samples calculated from gravimetric analysis (refer table 1) suggested that the grafting of HDA chains on CNTs increases with increase in reaction time. Xu et al. [16] reported similar results on a series of ODA grafted CNTs prepared by a three step procedure including acyl chlorination. Of the CNTs functionalized by two different procedures for the same time period, CNTs-1 showed higher amine content which may be due to the high reaction temperature.

Table 1. Correlation of weight % of HDA loaded on MWCNTS with reaction time

Time/h	Amine content/ wt%	
	CNTs-1	CNTs-2
3	10.3	2.8
6	24.8	6.9
12	52.4	12.2
24	61.5	21.7
92	89.7	51.4

Since preliminary studies showed a detrimental effect of higher amine content (> 35 wt% in CNTs) on the composite properties (results not given here) we chose CNTs with amine content in the range 20-25 % (refer table 1) for further investigations. So from now on, unless otherwise specified, CNTs-1 and CNTs-2 refer to CNTs functionalized for 6 and 24 h in a single and double step procedures, respectively.

Parts (a) and (b) of figure 1, respectively, show the FT-IR spectra of as received, functionalized CNTs from two different procedures and the corresponding PLA/CNT composites. The characteristic IR bands for CNTs are located between 1900 and 2100 cm^{-1} which are essentially the same for oxidized and functionalized CNTs indicating the preservation of graphite structure of CNTs even after oxidation and functionalization. The bands at 1740 cm^{-1} in the spectrum of O- CNTs (refer figure 1b) correspond to the C=O groups on the CNT surface formed during nitric acid oxidation. The spectral features of amine functionalized samples in CNTs-1 and CNTs-2, i.e. peaks around 2954–2850 cm^{-1} (stretching vibrations of the alkyl chain), 1500-1600 cm^{-1} (stretching vibrations of the amide) clearly indicate the incorporation of HDA to the CNTs surfaces,.[17-19] The FT-IR spectrum of PLA/CNT composite shows all the typical peaks of PLA at 1700-1760 and 500-1500 cm^{-1}, [20, 21] while retaining those for HDA and CNTs. This confirms the presence of functionalized CNTs in the polymer matrix. In the case of CNTs-2, the characteristic peak for CNTs is shifted to a higher wavenumber which is due to the covalent nature of the functionalization leading to amide formation. No such peak shift is observed in the case of CNTs-1 (refer figure1a) indicating that the HDA chains are adsorbed on the CNT surface without forming any covalent bonds.

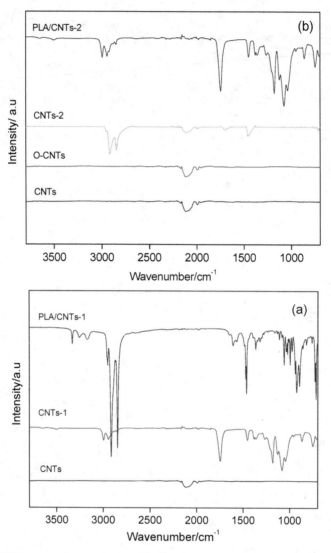

Figure1. FTIR spectra of as received, oxidized and amine functionalized CNTs and corresponding composite samples (a) one-step and (b) two-step functionalization

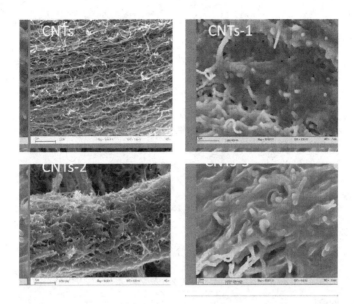

Figure 2. SEM images of as received and amine functionalized CNT samples. CNTs-1, 2, 3 represent CNTs functionalized by one, two and three step procedures respectively.

The degree of surface functionalization of CNTs was studied by SEM analysis and the results are given in figure 2. No visual difference between as received and oxidized CNTs was observed indicating that the acid treatment is non-destructive in nature (results not shown here). The SEM image of CNTs-1 show considerable amount of amine on the CNT surface when compared to CNTs-2 which is in agreement with the gravimetric analysis results for the amine content (refer table 1). For a comparison, SEM analysis was done on CNTs that are functionalized in a conventional three step procedure (referred to as CNTs-3) which involved CNT oxidation with concentrated HNO_3, refluxing with thionyl chloride for 8h followed by treatment with amine for 92 h. CNTs-1 and CNTs-3 are observed to have similar surface morphologies showing that the single step procedure is as effective as the three step procedure in functionalizing the CNT surface.

To find out the nature of bonding between the HDA chains with the outer graphene layer of the CNTs, XPS analyses was conducted. Figure 3a represents the XPS spectra of CNTs and CNTs-1 samples. The appearance of 'N 1s' peak at 400 eV, in CNTs-1 indicates the presence of amine groups on the CNT surface. The weak 'O 1s' and 'O KLL' peaks present in this case may be attributed to the presence of surface adsorbed oxygen in the sample. The characteristic 'C 1s' peak of CNTs-1 remains almost at the same position as that of CNTs, confirming that there is no covalent bond formation. On the other hand, the presence of covalently bonded amine groups on CNTs-2 could be identified by comparing the XPS spectra of CNTs, O-CNTs and CNTs-2 (refer figure 3b). The typical peaks observed were C1s at 284.05 eV, O1s around 527.7 and N1s at 395.6 eV for CNTs-2.[22-24] Moreover, the C1s lines of O-CNTs and CNTs-2 are significantly shifted to the lower binding energy values (~ 5 eV) indicating the successful amide formation on CNTs outer surfaces'.

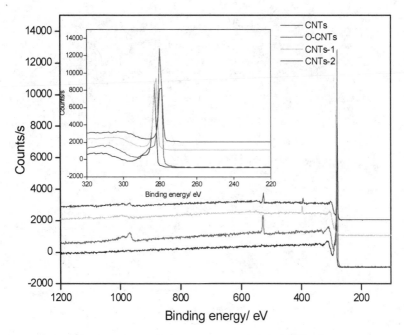

Figure 3. XPS scans of as received, oxidized and amine functionalized CNT samples. CNTs-1 and 2 represent CNTs functionalized by one and two step procedures respectively.

The PLA-CNT composite films were prepared with 1.5 wt% of functionalized CNTs (CNTs-1and CNTs-2). These films were subjected to TGA to measure their thermal stability and the resulting curves are shown in figure 4.

The onset of degradation temperature as measured from the intersection of the tangent of the initial point and inflection point is the same for both PLA and nanocomposite films. The PLA/CNT composites showed higher thermal stability when compared to the neat PLA which may be due to the homogenously distributed CNTs in the PLA and good interfacial interaction of CNTs with the polymer matrix. CNTs-2 are found to be slightly better in enhancing the thermal properties of the neat polymer which can be attributed to the presence of surface oxygen functionality on the CNTs which make them more compatible with PLA polymer.

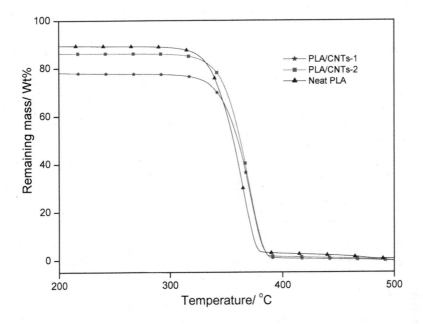

Figure 4. TGA curves for pure PLA and PLA-CNT composites

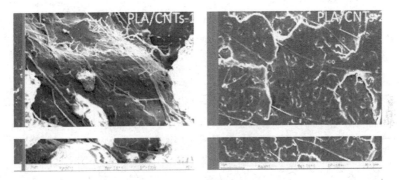

Figure 5. SEM images of fractured surface for PLA/CNT composites with 1.5 wt% of CNTs-1 and CNTs-2.

Figure 5 presents the SEM images of a typical fractured surface of the PLA-CNT nanocomposite films containing CNTs-1 and CNTs-2. It is evident that the CNTs are nicely distributed within the PLA matrix and offer good interfacial interaction. The long filament like morphology represents individual CNTs with preserved morphology. Therefore, solution mixing is found to be an efficient method to disperse individual CNTs by enhancing the interfacial adhesion with the PLA matrix. Such nicely distributed CNTs lead to enhanced thermal properties as revealed by TGA results.

CONCLUSIONS
In conclusion, amine functionalized CNTs were prepared by using two simple procedures. In both methods, the functionalized solids showed gradual grafting of HDA chains on CNTs surfaces' with increase in reaction time. Functionalization was demonstrated by FT-IR, SEM, and XPS analyses. Single step functionalization showed advantages over the two-step procedure in reducing the reaction time. XPS results indicated the formation of covalent bond through amide groups in the case of two step functionalization whereas the amine groups were just physisorbed on the CNT surface in one step functionalization. Solution mixing process resulted in a nice dispersion of CNTs with preserved aspect ratio within PLA matrix. The polymer nanocomposite thus prepared showed improved thermal properties when compared to the neat PLA.

ACKNOWLEDGEMENT
Authors thank Department of Science and Technology and Council for Scientific and Industrial Research, South Africa for the financial support.

REFERENCES
[1]B. I. Yakobson and R. E. Smalley, Fullerene Nanotubes: $C_{1,000,000}$ and Beyond, *Am. Sci.* **85**, 324 (1997).
[2]J. Liu, A. G. Rinzler, H. Dai, J. H. Hafner, R. K. Bradley, P. J. Boul, A. Lu, T. Liverson, K. Shelimov, C. B. Huffman, F. Rodriguez-Macias, Y. S. Shon, T. R. Lee, D. T. Colbert and R. E. Smalley, Fullerene Pipes, *Science* **280**, 1253 (1998).
[3]J. M. Bonard, T. Stora, J. P. Salvetat, F. Maier, T. Stöckli, C. Duschl, L. Forró, W.A. de Heer and A. Châtelain, Purification and Size-Selection of Carbon Nanotubes, *Adv. Mater.* **9**, 827 (1997).
[4]B. Vigolo, A. Pe´nicaud, C. Coulon, C. Sauder, R. Pailler, C. Journet, P. Bernier and P. Poulin, Macroscopic Fibers and Ribbons of Oriented Carbon Nanotubes, *Science* **290**, 1331 (2000).
[5]M. F. Islam, E. Rojas, D. M. Bergey, A. T. Johnson and A. G. Yodh, High Weight Fraction Surfactant Solubilization of Single-Wall Carbon Nanotubes in Water, *Nano Lett.* **3**, 269 (2003).
[6]V. C. Moore, M. S. Strano, E. H. Haroz, R. H. Hauge and R. E. Smalley, Individually Suspended Single-Walled Carbon Nanotubes in Various Surfactants, *Nano Lett.* **3**, 1379 (2003).
[7]A. B. Dalton, C. Stephan, J. N. Coleman, B. McCarthy, P. M. Ajayan, S. Lefrant, P. Bernier, W. J. Blau and H. J. Byrne, Selective Interaction of a Semiconjugated Organic Polymer with Single-Wall Nanotubes, *J. Phys. Chem. B* **104**, 10012 (2000).
[8]A. Star, J. Fraser Stoddart, D. Steuerman, M. Diehl, A. Boukai, E. W. Wong, X. Yang, S. W. Chung, H. Choi and J. R. Heath, Preparation and Properties of Polymer-wrapped Single-Walled Carbon Nanotubes, *Angew. Chem., Int. Ed.* **40**, 1721 (2001).
[9]M. J. O'Connell, P. Boul, L. M. Ericson, C. Huffman, Y. Wang, E. Haroz, C. Kuper, J. Tour, K. D. Ausman and R. E. Smalley, Reversible Water-solubilization of Single-Walled Carbon Nanotubes by Polymer Wrapping,*Chem. Phys. Lett.* **342**, 265 (2001).

[10]J. Chen, H. Liu, W. A. Weimer, M. D. Halls, D. H. Waldeck and G. C. J. Walker, Noncovalent Engineering of Carbon Nanotube Surfaces by Rigid, Functional Conjugated Polymers, *Am. Chem. Soc.* **124**, 9034 (2002).

[11]S. Ramesh, L. M. Ericson, V. A. Davis, R. K. Saini, C. Kittrell, M. Pasquali, W. E. Billups, W. W. Adams, R. H. Hauge and R. E. Smalley, Dissolution of Pristine Single Walled Carbon Nanotubes in Superacids by Direct Protonation, *J. Phys. Chem. B* **108**, 8794 (2004).

[12]S. S. Wong, E. Joselevich, A. T. Woolley, C. L. Cheung and C. M. Lieber, *Nature* **394**, 52 (1998).

[13]J. Chen, A. M. Rao, S. Lyuksyutov, M. E. Itkis, M. A. Hamon, H. Hu, R. W. Cohn, P. C. Eklund, D. T. Colbert, R. E. Smalley and □R. C. Haddon, Dissolution of Full-Length Single-Walled Carbon Nanotubes, *J. Phys. Chem. B* **105**, 2525 (2001).

[14]Y. Qin, L. Liu, J. Shi, W. Wu, J. Zhang, Z. X. Guo, Y. Li and D. Zhu, Large scale Preparation of Solubilized Carbon Nanotubes, Chem. Mater, **15**, 3256 (2003).

[15]M. A. Hamon, H. Hu, P. Bhowmik, S. Niyogi, B. Zhao, M. E. Itkis and R. C. Haddon, End-group and Defect analysis of Soluble Single-Walled Carbon Nanotubes , *Chem. Phys. Lett.* **347**, 8 (2003).

[16]M. Xu, Q. Huang, Q. Chen, P. Guo and Z. Sun, Synthesis and Characterization of Octadecylamine Grafted Multi-walled Carbon Nanotubes, *Chem. Phys. Lett.* **375**, 598 (2005).

[17]H.-C. Kuan, C.-C.M. Ma,W.-P. Chang, S.-M. Yuen, H.-H.Wu and T.-M. Lee, Synthesis, Thermal, Mechanical and Rheological properties of Multiwall Carbon Nanotube/Water borne Polyurethane Nanocomposite , *Compos. Sci. Technol.* **65**, 1703 (2005).

[18]J.L. Stevens, A.Y. Huang, H. Peng, I.W. Chiang, V.N. Khabashesku and J. L. Margrave, Sidewall Amino-Functionalization of Single-Walled Carbon Nanotubes through Fluorination and Subsequent Reactions with Terminal Diamines, *Nano Lett.* **3**, 331 (2003).

[19]H. Peng, L.B. Alemany, J.L. Margrave and V.N. Khabashesku, Sidewall Carboxylic Acid Functionalization of Single-Walled Carbon Nanotubes, *J. Am. Chem.Soc.* **125**, 15174 (2003).

[20]N. Ignjatovic, V. Savic, S. Najman, M. Plavsic and D. Uskokovic, A Study of Hap/PLLA composites as a substitute for bone powder using FT-IR Spectroscopy, *Biomater.* **22**, 571 (2001).

[21]Y. Xiao, Y. Xu, J. Lu, X. Zhu, H. Fan and X. Zhang, Preparation and Characterization of Collagen-modified Polylactide Microparticles , *Mater. Lett.* **61**, 2601 (2007).

[22]G. Gabriel, G. Sauthier, J. Fraxedas, M. Moreno-Mañas, M. T. Martinez, C. Miravitlles and J. Casabo, Preparation and Characterisation of Single-Walled Carbon Nanotubes Functionalised with Amines , *Carbon* **44**, 1891 (2006).

[23]J. Shen, W. Huang, L. Wu, Y. Hu, M. Ye, Study on amino-functionalized multiwalled carbon nanotubes, *Mater. Sci. Eng. A* **464**, 151 (2007).

[24]X. Li and M. R. Coleman, Functionalization of Carbon Nanofibers with Diamine and Polyimide Oligmer, *Carbon* **46**, 1115 (2008).

THE EFFECT OF SURFACE FUNCTIONALIZED CARBON NANOTUBES ON THE
MORPHOLOGY, AS WELL AS THERMAL, THERMOMECHANICAL, AND
CRYSTALLIZATION PROPERTIES OF POLYLACTIDE

James Ramontja[1, 2], Suprakas Sinha Ray[1, 2], Sreejarani K. Pillai[1] and Adriaan S Luyt[2]
[1]DST/CSIR Nanotechnology Innovation Centre, National Centre for Nano-Structured Materials,
Council for Scientific and Industrial Research, Pretoria 001, South Africa.
[2]Department of Chemistry, University of the Free State (Qwaqwa Campus), Phuthaditjhaba,
South Africa.

ABSTRACT
    This paper discusses various properties of Poly(lactide) upon nanocomposite formation
with functionalized multiwalled carbon nanotubes (f-MWCNTs). The composite was prepared
through melt extrusion technique. Functionalization of carbon nanotubes and possible interaction
with PLA chains was investigated through attenuated total reflectance (ATR) Fourier
transformed-infrared (FT-IR) and Raman spectroscopies. Scanning electron microscope (SEM)
and polarized optical microscope (POM- in melt state) also revealed homogenous dispersion of
f-MWCNTs in the PLA matrix with some agglomerates. Melting and crystallization phenomena
of the nanocomposite studied through differential scanning calorimeter (DSC), wide angle X-ray
scattering (WAXS), and POM show that f-MWCNTs facilitates nucleation and crystal growth of
PLA matrix significantly. Thermogravimetric analyses (TGA) reveal that overall thermal
stability of PLA matrix improves slightly upon the nanocomposite formation. Thermomechanical
analyses also reveal a significant increase in modulus of the nanocomposite at room temperature,
which drops suddenly across glass transition temperature. This is an indication of plasticization
effect.

INTRODUCTION
    Over the last two decades, the world has embarked on a massive research in the field of
biodegradable and biocompatible polymers, both for medical and ecological applications.[1] One
such polymer is polylactide (PLA), for it is readily biodegradable and is made from agricultural
sources[2]. PLA has potential medical applications such as tissue culture, surgical implants,
restorable sutures, wound closure, and controlled-release systems.[3-5] Polylactide is not only
biocompatible but also bioresorbable. When implanted in living organisms including human
body, it is hydrolyzed to its constituent α-hydroxy acid which is eliminated by general metabolic
pathways[6]. However, for biomedical applications, neat PLA might not be suitable for high load
bearing applications[7], which intrigued the need to incorporate the reinforcements such as
oriented PLA fibers. Nanocomposites represent an exceptional case of composites in which
interfacial relationship between two phases is maximized.
    In recent years, a significant amount of work has been done on the preparation and
characterization of polymer nanocomposites based on nanoclays such as montmorillinite,
saponite, and synthetic mica.[8-14] These fillers moderately improved the mechanical and physical
properties of the neat polymer matrices even though their amounts were small (~5 wt.%). The
main reason for these improved properties in the case of the clay-containing polymer
nanocomposites is the presence of interfacial interactions as opposed to the conventional
composites.

Currently a number of researchers are focusing on the preparation and characterization of functionalized carbon nanotube containing polymer nanocomposites.[15-27] This is because CNTs have superior mechanical properties such as extraordinary high strength, high modulus, excellent electrical conductivity along with their thermal conductivity and stability, and the low density associated with high aspect ratio compared to other nano-fillers.[28,29] However, the effective utilization of CNTs has not being realized due to difficulties in producing CNT/polymer nanocomposites with homogeneously well-dispersed CNTs.[30] Due to intrinsic van der Waals interactions,[31] the as received CNTs tend to aggregate and entangle together spontaneously when blended directly with polymers. With poor dispersion, the active surface area for polymer/CNT surface interaction will not increase sufficiently and as a result very small amount of stress will be transferred between CNT filler and polymer matrix. It has been reported that chemical modification on the surface of CNTs improves their dispersion on polymer matrices.[32-36] This chapter summarizes various properties of a PLA composite containing 0.5 wt.% of f-MWCNTs. The f-MWCNTs used in this work contain ~20% (determined gravimetrically) of hexadecylamine (HDA).

EXPERIMENTAL

Materials

PLA (weight average molecular weight = 188k g.mol^{-1}) with a D-lactide content of 1.1–1.7% was obtained from Unitika Co. Ltd, Japan. Prior to use, PLA was dried at 80 °C for 2 days under vacuum. The CNTs (here multi-walled CNTs) used in this study were synthesized by chemical vapour deposition (inner diameter ~10 nm; outer diameter ~ 20 nm; average length ~500 μm, measured by transmission electron microscopy) and 95% pure (measured by energy dispersive X-ray spectroscopy). Hexadecylamine (HDA), chloroform, and ethanol were purchased from Sigma-Aldrich and used as received.

In a typical functionalization process, a mixture of 1g CNTs and 5g HDA was taken in a conical flask and heated at 180°C for 6 h in an oil bath. After cooling to room temperature, the excess of HDA was removed from the reaction mixture by washing with ethanol several times. The black solid was then collected by Nylon membrane filtration (0.45 μm pore size) and dried at 110°C overnight to get a constant weight. The increased weight of the CNTs was ~20%, determined gravimetrically. This results imply that the amount of HDA surfactant content is ~20 wt.-%

Preparation of Nanocomposites

For the preparation of PLA/MWCNT nanocomposite, f-MWCNTs (0.5 wt.-%, powder form) and PLA (pellet form) were first dry mixed in a polyethylene bottle. The mixture was then extruded using co-rotating twin-screw mini-extruder (bench-top Haake Minilab II, Thermo Scientific) operated at 180°C (screw speed = 30 rpm, time = 5 min) to yield black nanocomposite strands. These strands were chopped into pieces and stacked between two metal plates and compression molded by pressing with 2 MPa pressure at 180 °C for 2 min. Neat PLA and nanocomposite samples were annealed at 110 °C under vacuum prior to all characterizations and property measurements.

Characterization and property measurement

The functionalization and the presence of f-MWCNTs was confirmed through the attenuated total reflectance (ATR) Fourier-transform infrared (FT-IR) using Perkin Elmer Spectrum 100 instrument at a resolution of 4.0 cm^{-1}. Raman spectroscopy studies were employed using a lab Raman system, Jobin-Yvon Horiba T64000 Spectroscopy, equipped with an Olympus

BX-40 microscope. The excitation wavelength was 514.5 nm with an energy setting of 1.2 mW from a Coherent Innova model 308 argon ion laser. The morphology of the freeze fractured surface of the composite was analyzed using Carl Zeiss SMT Neon 40, Cross Beam Series FIB-SEM in SEM mode, with an acceleration voltage of 2 kV. The spherulitic growth behaviour and the degree of dispersing in molten state of neat PLA and its composite were studied with a Carl Zeiss Imager Z1M polarized optical microscope (POM). Samples were heated to 190°C at a heating rate of 20°C.min^{-1}, held at that temperature for 5 min, and then the pictures were taken.

The melting and glass transition temperatures as well as crystallinity of the PLA matrix before and after nanocomposite formation were studied with a DSC instrument (model: TA Q2000) under constant nitrogen flow of 50 mL.min^{-1} and a heating rate of 20 °C.min^{-1}. WAXS experiments of the PLA and nanocomposite samples were carried out in an Anton Paar SAXS instrument operated at 40 kV and 50 mA with line collimation geometry. The radiation used was a Ni filtered CuK$_\alpha$ radiation of wavelength 0.154 nm (PAN Analytical X-ray source). Thermogravimetric analyses of both PLA and the nanocomposite samples were carried out on a TGA Q500 (TA Instruments) at a heating rate of 10 °C.min^{-1} under thermo-oxidative conditions, from ambient temperature to 650 °C. The dynamic mechanical properties of neat PLA and its composite samples were determined using an Anton Paar-Physica MCR501 Rheometer in the tension-torsion mode. The temperature dependence of the storage modulus (G′) and tan δ of neat PLA and composite samples, were measured at a constant frequency (ν) of 6.28 rad.s^{-1} with the strain amplitude of 0.02% (selected after a series of strain sweep tests at different temperatures to determine the linear region) and in the temperature range of -20 to 160 °C at a heating rate of 2 °C.min^{-1}.

RESULTS AND DISCUSSION

Attenuated total reflectance fourier-transform infrared (ATR-FTIR) spectroscopy

Figure 1 shows the ATR-FTIR spectra of neat PLA, the f-MWCNTs and the PLA/f-MWCNTs composite. The spectrum of the composite shows the characteristic peaks of PLA and the f-MWCNTs. The broad peak in the spectrum of the f-MWCNTs represents the N-H stretching of HDA. This broad peak also appears in the spectrum of the composite. The peak at 1592 cm^{-1} (indicated by *) in the spectrum of the f-MWCNTs represents the primary amine N-H deformation of HDA. This peak is also observed in the spectrum of the composite at 1645 cm^{-1} (also indicated by *). These results confirm the presence of f-MWCNTs in the composite. However, it is difficult to establish whether there is a possible interfacial interaction between the PLA and the HDA chains, because we could not get a clear peak of the N-H stretching in both the f-MWCNTs and the composite spectra.

Raman spectroscopy

Raman spectroscopy was used to verify the presence of possible interfacial interactions between f-MWCNTs and the matrix of PLA. Figure 2 shows the Raman spectra of the f-MWCNTs and the corresponding nanocomposite of PLA. It can be seen from the spectra that there is a small shift in the characteristic D-band and a quite significant shift in the G-band of the f-MWCNTs to higher wavenumbers in the case of the nanocomposite. This indicates the presence of interfacial interactions between the PLA chains and the f-MWCNTs surfaces. It can also be seen that the characteristic peak of the PLA matrix (appearing at 1450 cm^{-1} for neat PLA,

[26]) moves toward higher a wavenumber of 1453 cm[-1]. This observation further confirms the presence of some interactions between the PLA matrix and the f-MWCNT surfaces.

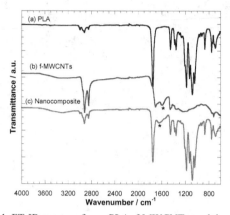

Figure 1. FT-IR spectra of pure PLA, f-MWCNTs, and the nanocomposite.

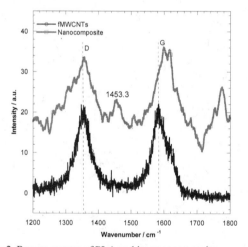

Figure 2. Raman spectra of PLA and its nanocomposite.

Scanning electron microscopy

The dispersion of the f-MWCNTs in the PLA matrix was studied using a scanning electron microscope (SEM) operated at an accelerated voltage of 2 kV. figure 3 (a) represents the SEM image of the freeze fractured surface of the PLA/f-MWCNT nanocomposite. The polymer matrix surface with some white spots is clearly seen. Two areas, with and without white spots, were selected and magnified. They are shown in figure 3(b & c). In these pictures a fairly good dispersion of CNTs can be seen, but the white spots are clearly the result of agglomeration of f-MWCNTs in the PLA matrix at a micron scale level. Agglomeration of MWCNTs in the PLA matrix suggests that part of the surface area of CNTs could not be accessed during functionalization by HDA. This is due to the intrinsic van der Waals forces keeping the MWCNTs together as bundles. Based on these observations, it can be concluded that the homogenous dispersion of the f-MWCNTs is the result of improved interaction between the PLA matrix and the HDA chains on the surface of the MWCNTs, as established in Raman results.

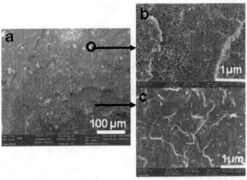

Figure 3. Scanning electron microscopy images of a nanocomposite containing 0.5 wt.% f-MWCNTs with two selected spots at different magnifications.

Polarized optical microscopy

To further verify the good dispersion of the f-MWCNTs in the PLA matrix, the composite was investigated through an optical microscopy at 190 °C where PLA was in the molten state. These results are presented in figure 4. The dark spots represent agglomerates of CNTs. This image clearly shows that there is an improved dispersion of f-MWCNTs with few agglomerates at micron scale, as shown by the dark spots. Again, this results support the SEM results.

Figure 4. Optical microscopic image of the PLA/f-MWCNTs nanocomposite taken at 190 °C in the transmittance mode. This is the representative image of images taken from five different positions.

The effect of cooling rate on the non-isothermal crystallization behaviour of PLA

To study the influence of cooling rates on the non-isothermal crystallization behaviour of PLA, the samples were heated to 190 °C at a heating rate of 20 °C.min[-1], kept at this temperature for 5 min, and then cooled down to -20 °C at different cooling rates. The cooling curves of pure PLA and its composite during non-isothermal crystallization from their melts at five different cooling rates are shown in figure 5. In the case of neat PLA, a broad peak is observed when the sample was cooled from the melt at a rate of 0.5 °C.min[-1]. With an increase in cooling rate to 1 °C.min[-1], a peak with a shoulder peak appears and shifts towards lower temperatures. The peak shoulders indicate a continuous change of enthalpy. It is clear that at cooling rates higher than 5 °C.min[-1], it is very difficult for the PLA matrix to fully crystallize and the polymer stays in a super-cooled state. The crystallization peak shifts to lower temperatures as the cooling rate is increased is a natural observation, because it is difficult for the polymer chains to crystallize at faster cooling rates. A small crystallization peak appears at 126 °C for the composite when the cooling rate from the melt is 0.5 °C.min[-1]. It is further observed that this peak does not clearly show the double thermal event that was observed in the case of the neat PLA. This peak also shifts to lower temperatures as the cooling rate increases to 1 °C.min[-1]. A further increase in the cooling rates to 5 °C.min[-1] also shows the presence of a double peak as in the case of PLA. The crystallization peaks for the nanocomposite, for all the investigated cooling rates, are more intense and better resolved than those for neat PLA. What is more interesting is that even at a faster cooling rate of 10 °C.min[-1], the nanocomposite is still able to crystallize. Based on the observations above, it can be concluded that f-MWCNTs act as nucleating agents for the crystallization of the PLA matrix.

To confirm the nucleating effect of the f-MWCNTs during non-isothermal crystallization, the samples were investigated through POM. For the POM measurements, a cooling rate of 10 °C.min[-1] was selected because during injection moulding the cooling rates are usually very fast. The POM images of the PLA and its nanocomposite, taken at 130 °C during isothermal crystallization from their melt, are shown in figure 6. The images show large spherulites for the neat PLA sample, but much smaller and more densely packed crystallites for the nanocomposite. This observation indicates that the f-MWCNT nanoparticles formed nucleating sites for the formation of small spherulites in the nanocomposite.

Effect of cooling rates on melting behaviour of PLA

In order to study the effect of cooling rates on the melting behaviour, PLA and its nanocomposite were heated from -20 to 190 °C at 20 °C.min[-1] as soon as the cooling was finished. These heating curves are presented in figure 7, and the DSC data are summarized in Table 1. It can be seen that PLA only shows cold crystallization peaks and two melting peaks when the cooling rates were 5 and 10 °C min[-1]. The composite also shows cold crystallization peaks at the same cooling rates, but single melting peaks. This observation indicates that the crystallization of PLA chains was not completed during cooling at the faster cooling rates, and the crystallization process continued during heating. The double melting peaks indicate the presence of different types of crystals with different stabilities. Nam et al.[27] also suggested that the double melting peaks of PLA may be due to the presence of less perfect crystals having enough time to melt and rearrange into crystals with higher structural perfection, which re-melted at higher temperatures during heating in the DSC. However, when the cooling rates were 0.5, 1, and 2 °C.min[-1], no cold crystallization peak was observed for both PLA and its nanocomposite. This indicates that the crystallization of PLA chains was completed at slower cooling rates during the non-isothermal cooling process. Single melting peaks were observed when the cooling rate was 0.5 °C.min[-1] for both samples. In brief, the nanocomposite shows two distinct melting peaks when the cooling rate was 2 °C.min[-1] in comparison to the neat polymer. This is an indication that the nucleation effect of f-MWCNTs in the polymer matrix assisted in the formation of more perfect crystals.

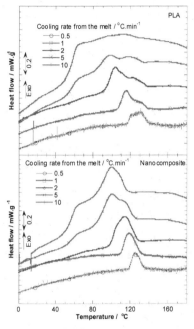

Figure 5. DSC heating curve of PLA and its nanocomposite after non-isothermal crystallization at different cooling rates.

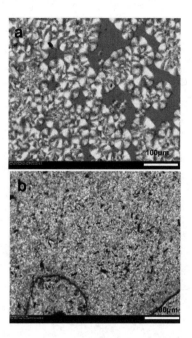

Figure 6. Polarized optical micrographs of (a) neat PLA and (b) the PLA/f-MWCNT nanocomposite. Both samples were crystallized at 130 °C from their melts.

By integrating the area under the endothermic region of the DSC curves, and by subtracting the extra heat absorbed by the crystallites formed during cold crystallization, the melting enthalpy (ΔH_m) of all the samples was calculated, and at the same time the degree of crystallinity (χ_c) was estimated by considering the melting enthalpy of 100% crystalline PLA as 93 J.g[-1].[21] The χ_c data in Table 1 show that the overall crystallinity of PLA was reduced when 0.5 wt.% f-MWCNTs was added. A decrease in overall crystallinity may be as a result of two factors: MWCNT agglomerates acting as active nucleation sites and at the same time, the non-agglomerated sites inhibiting mobility of the polymer chains. Because of the well-dispersed f-MWCNTs crystal growth was inhibited, thus leading to a decrease of crystallinity.

Table 1   Cooling rate dependence of the melting enthalpy from two melting peaks of the PLA and the composite

Sample	Cooling rate	Melting enthalpy / J g^{-1} [a]	% crystallinity[b]
PLA	0.5	53.0	57.0
	1	44.4	47.8
	2	41.1	44.2
	5	37.6	40.4
	10	37.6	40.4
Nanocomposite	0.5	48.0	51.6
	1	40.1	43.1
	2	39.1	42.0
	5	38.8	41.7
	10	33.9	36.5

[a] The total melting enthalpy of PLA evaluated by integration of the area under the endothermic peaks from the heating scans after non-isothermal crystallization.
[b] Calculated using the melting enthalpy of 100% crystalline PLA, 93 J g^{-1}.[37]

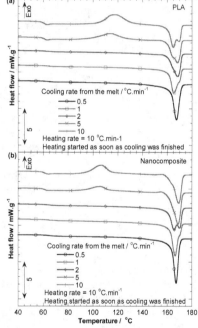

Figure 7. DSC heating curves of PLA and the nanocomposite after non-isothermal crystallization at five different cooling rates.

Temperature modulated DSC

To separate the heat capacity and kinetically related components during cold crystallization and subsequent melting of neat PLA and its nanocomposite, TMDSC of melt quenched samples were done. TMDSC allows us to see whether any re-crystallization process occurs as soon as PLA begins to melt. This has been used to confirm the presence of melting, re-crystallization, and re-melting processes. Figure 8 illustrates the TMDSC curves of (a) neat PLA and (b) its nanocomposite during the second heating. The samples were first equilibrated at -20 °C for 30 min, and then heated to 190 °C at a rate of 2 °C.min^{-1}, kept at that temperature for 5 min. to destroy any previous thermal history, and cooled to -20 °C at a rate of 2 °C.min^{-1}. TMDSC was started as soon as the cooling was finished. For both samples the total heat flow (middle curve) is separated into well defined reversible heat flow (bottom curve) and non-reversible heat flow (top curve). For neat PLA, the following behaviour is observed: two melting signals on the reversible heat flow curve are accompanied by the subsequent re-crystallization on the non-reversible heat flow curve, with the total heat flow curve showing only the melting peaks. This observation may be due to the partial melting and perfection of different crystals at temperatures before their final melting. For the nanocomposite, two melting peaks are observed for all the heat flow curves with no apparent re-crystallization. What is more notable is that the two melting peaks of the nanocomposite on the reversible heat flow curve are now distinct in comparison with the peaks for the neat polymer. This indicates the presence of different forms of crystals with different thermal stabilities. Another interesting feature is that TMDSC enabled us to see partial re-crystallization occurring in the neat polymer, which is absent in the composite.

To estimate the percent crystallinity (χ_c) of the samples, we took the enthalpy of melting (ΔH_f) from the reversible heat flow curve, divided this value by the enthalpy of a 100% crystalline polymer (ΔH_f for 100 % crystalline PLA is 93 J.g^{-1} [21]), and multiplied the answer by 100%. The data is reported in table 2. These values indicate that the crystallinity of the PLA matrix decreased in the presence of the f-MWCNTs.

Table 2 TMDSC data for PLA and its nanocomposite.

Sample	Total			Reversible				Non-reversible			χ_c
	T_{m1} °C	T_{m2} °C	ΔH_f J.g^{-1}	T_g °C	T_{m1} °C	T_{m2} °C	ΔH_f J.g^{-1}	ΔH_c J.g^{-1}	ΔH_f J.g^{-1}	T_m °C	%
PLA	169.9	-	43.54	62.2	164.2	169.8	30.1	8.9	26.7	169.9	19.1
Composite	165.3	171.2	44.94	62.0	165.3	171.2	26.9	1.7	24.3	171.0	24.3

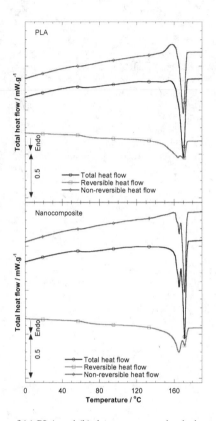

Figure 8. TMDSC curves of (a) PLA and (b) the nanocomposite during second heating.

Wide angle X-ray scattering

To study the presence of different PLA crystals and their modification, WAXS of the neat PLA and nanocomposite samples were performed. The measurements were taken from room temperature to the melting temperature, and then back to room temperature. The samples were kept at each temperature for 5 minutes, including 1 minute exposure to the X-rays. Figure 9 shows the one-dimensional WAXS patterns of PLA and the nanocomposite obtained under these conditions. Overall, there is no sign of the modification of existing crystals or the formation of new crystals. The notable observation is when both samples were cooled from their melts. It is clear that it is very difficult for PLA to crystallize during cooling. However, crystals are formed in the presence of f-MWCNTs as shown by the fully resolved peaks in the spectra of the nanocomposite. Again, this supports the nucleation effect of f-MWCNTs in the polymer matrix. However, a very small peak is observed on the spectra of both samples at around $2\Theta = 22.5°$. This observation suggests the growth of another type of crystal.

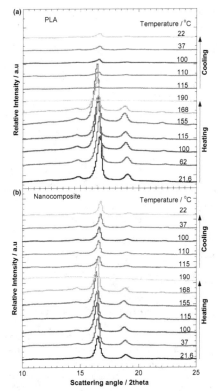

Figure 9. Temperature dependence wide-angle X-ray scattering patterns of (a) neat PLA and (b) the nanocomposite samples during both heating and cooling cycles.

Thermogravimetric analysis

This section discusses the thermal stabilities of neat PLA and the nanocomposite in a thermo-oxidative environment. The TGA and the first dTGA curves of neat PLA and the nanocomposite obtained under oxygen flow are presented in figure 10. The dTGA are presented because they more clearly show the difference in thermal stabilities between the samples. Both samples show a one-step decomposition. The thermal stability of the nanocomposite is higher than that of the neat PLA. This improvement can be attributed to the fairly homogenous dispersion of the f-MWCNTs. The thermal stability of the nanocomposite may also be due to the higher thermal stability of the CNTs in comparison to that of PLA. The dTGA peak of the nanocomposite shifts to a higher temperature compared to that of the neat PLA sample. This is also an indication of the improvement in thermal stability of PLA in the presence of the f-MWCNTs.

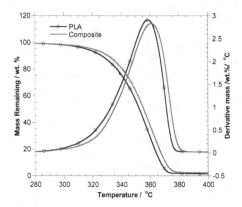

Figure 10. TGA and derivative TGA curves of PLA and the nanocomposite under oxygen flow at a heating rate of 10 °C min^{-1}.

Dynamic mechanical analysis

DMA generally reveals the amount of energy stored in the nanocomposite as elastic energy, and the amount of energy dissipated during mechanical strain, which strongly depends on the geometrical characteristics and the level of dispersion of the filler in the matrix. It also depends on the degree of interaction between the matrix and the filler.[38] Figure 11 (a and b) represents the storage modulus (G') and the damping factor (tan δ) curves for PLA and the nanocomposite, respectively. The damping factor provides information on the relative contributions of the viscous and elastic components of the viscoelastic material. Figure 11a shows three phenomena: (1) from 0-50 °C, there is an increase in modulus. This is because both samples are stiff because there is not yet chain mobility, but the nanocomposite is stiffer due to the presence stiff f-MWCNTs; (2) from 50-80 °C, there is a sudden drop of modulus because the chains of the surfactant (HDA) exhibits a plasticizing effect on the polymer matrix just below and above the glass transition temperature; (3) from 80-160 °C, there is a slight improvement of modulus because the presence of fairly homogenously dispersed f-MWCNTs inhibits the PLA chain mobility. Figure 11b clearly indicates that there is a decrease in the glass transition temperature from 74 to 71 °C. This supports the observation of a plasticizing effect of the HDA chains in the PLA matrix. From these observations it may be concluded that the fairly homogenously dispersed f-MWCNTs in the PLA matrix improved the storage modulus below and above the glass transition temperature. Also, the f-MWCNTs acted as a plasticizer of the PLA matrix at around the glass transition temperature.

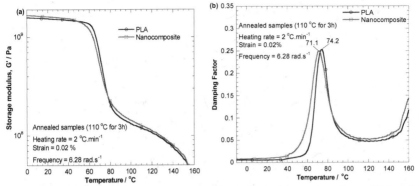

Figure 11. Temperature dependence of dynamic mechanical properties of neat PLA and its nanocomposite: (a) storage modulus and (b) damping factor.

CONCLUSIONS

This paper discussed the morphology, thermal, and thermomechanical properties of a PLA nanocomposite containing 0.5 wt.% of f-MWCNTs (with an amine content of ~20 %). The SEM and POM (of the samples in the molten state) results confirm the homogenous dispersion of f-MWCNTs in the PLA matrix, with some micro-agglomeration. The POM results also show the formation of much smaller PLA crystallites in the presence of f-MWCNTs. The f-MWCNTs were found to play a nucleation role in the crystallization of PLA, as observed from the DSC, SEM, and WAXS results. The DMA and TGA results show that the presence of f-MWCNTs had only a slight influence on the thermomechanical properties and thermal stability of the PLA. FTIR and Raman spectroscopy confirmed the functionalization of the MWCNTs, and the presence of facial interaction between f-MWCNTs and the PLA matrix.

ACKNOWLEDGEMENTS

We wish to thank the CSIR executive and the DST, South Africa, for financial support.

REFERENCES

[1] Y. Ikada and H. Tsuji, Biodegradable polyesters for medical and ecological applications, *Macomol. Rapid Commun.*, **21**, 117-32 (2000).
[2] P. Gruber and M. O'Brien, Polyesters III Applications and Commercial Products, Biopolymers edited by Y. Doi and A. Steinbuchel, Wiley-VCH, Weinheim, vol., **4**, 235 (2002).
[3] V. Krikorian and D. J. Pochan, Poly (L-Lactic Acid)/Layered Silicate Nanocomposite: Fabrication, Characterization, and Properties, *Chem. Mater*, **15**, 4317-24 (2003).
[4] R. A. Jain, The manufacturing techniques of various drug loaded biodegradable poly(lactide-*co*-glycolide) (PLGA) devices, *Biomaterials.*, **21**, 2475-90 (2000).
[5] K. R. Kamath and K. Park, Biodegradable hydrogels in drug delivery, *Adv. Drug Delivery Rev.*, **11**, 59-84 (1993).
[6] (a) E. J Frazza, E. E Schmitt, A new absorbable suture, *J. Biomed. Mater. Res. Symp.*, **1**, 43-58 (1971). (b) M Vert, Biomedical Polymers from Chiral Lactides and Functional. Lactones: Properties and Applications, *Makromol. Chem., Makromol. Symp.*, **6**, 109 (1986).

[7]N. C. Bleach, S. N. Nazhat, K. E. Tanner, M. Kellomaki, and P. Tormala, Effect of filler content on mechanical and dynamic mechanical properties of particulate biphasic calcium phosphate polylactide composites *Biomaterials.*, **23**, 1579-85 (2002).

[8]S. Sinha Ray, P. Maiti, M. Okamoto, K. Yamada, and K. Ueda. New polylactide/layered silicate nanocomposites. I. Preparation, characterization and properties. *Macromol.*, **35**, 3104–10 (2002).

[9]S. Sinha Ray, K. Yamada, M. Okamoto, A. Ogami, and K. Ueda, New polylactide/layered silicate nanocomposites. 3. High performance biodegradable materials, *Chem. Mater.*, **15**, 1456–65 (2003).

[10]S. Sinha Ray, K. Yamada, M. Okamoto, and K. Ueda. Biodegradable polylactide/montmorillonite nanocomposites, *J. Nanosci. Nanotech.*, **3**, 503-10 (2003).

[11]C.R. Tseng, J.Y. Wu, Y.H. Lee, and F.C. Chang, Preparation and crystallization behaviour of syndiotactic polystyrene-clay nanocomposites, *Polymer.*, **42**, 10063-70 (2001).

[12]R.A. Vaia, H. Ishii, E.P. Giannelis. Synthesis and properties of two-dimensional nanostructures by direct intercalation of polymer melts in layered silicates, *Chem Mater.*, **5**, 1694-96 (1993).

[13]M. Okamoto, S. Morita, and T. Kotaka. Dispersed structure and ionic conductivity of smectic clay/polymer nanocomposites, *Polymer.*, **42**, 2685-88 (2001).

[14]B. Lepoitevin, N. Pantoustier, M. Alexandre, C. Calberg, R. Jerome, and P. Dubois, Polyester layered silicate nanohybrids by controlled grafting polymerization, *J. Mater Chem.*, **12**, 3528-32 (2002).

[15]F.T. Fisher, R.D. Bradshaw, and L.C. Brinson, Effects of nanotube waviness on the modulus of nanotube-reinforced polymers, *Appl Phys Lett.*, **80**, 4647-49 (2002).

[16]Z. Yao, N. Braidy, G.A. Botton, and A. Adronov, Polymerization from the surface of single-walled carbon nanotubes – Preparation and characterization of nanocomposites *J. Am. Chem. Soc.*, **125**, 16015–24 (2003).

[17]S. Qin, D. Qin, W.T. Ford, D.E. Resasco, and J.E Herrera, Functionalization of single-walled carbon nanotubes with polystyrene via grafting to and grafting from methods. *Macromolecules.*, **37**, 752–57 (2004).

[18]J. Chen, R. Ramasubramaniam, C. Xue, and H. Liu, A versatile molecular engineering approach to simultaneously enhanced, multifunctional carbon-nanotube–polymer composites, *Adv. Funct. Mater.*, **16**, 114-19 (2006).

[19]P. Calvert. Nanotube composites: A recipe for strength, *Nature.*, 399, 210-11 (1999).

[20]H.S. Kim, B.H. Park, J.S. Yoon, and H.J. Jin. Thermal and electrical properties of poly(L-lactide)-graft-multiwalled carbon nanotube composites, *Eur. Polym. J.*, **43**, 1729–35 (2007).

[21]Y.T. Shieh and G.L. Liu, Effects of carbon nanotubes on crystallization and melting behaviour of poly(L-lactide) via DSC and TMDSC studies, *J. Polym. Scie. B Polym. Phys.*, **45**, 1870–81 (2007).

[22]G.X. Chen, H.S. Kim, B.H. Park, and J.S. Yoon, Synthesis of poly(L-lactide)-functionalized multiwalled carbon nanotubes by ring-opening polymerization. *Macromol. Chem. Phys.*, **208**, 389–98 (2007).

[23]H. Tsuji, Y. Kawashima, H. Takikawa, and S. Tanaka, Poly(L-lactide)/nano-structured carbon composites: Conductivity, thermal properties, crystallization, and biodegradation, *Polymer.*, **48**, 4213–25 (2007).

[24]G.X. Chen and H. Shimizu, Multiwalled carbon nanotubes grafted with polyhedral

oligomeric silsesquioxane and its dispersion in poly(L-lactide) matrix., *Polymer.*, **49**, 943–91 (2008).

[25]D. Wu, L. Wu, M. Zhang, and Y. Zhao, Viscoelasticity and thermal stability of polylactide composites with various functionalized carbon nanotubes, *Polym Degrad. Stab.*, **93**, 1577–84 (2008).

[26]J. Ramontja, S. Sinha Ray, S.K. Pillai, and A.S. Luyt, High-performance carbon nanotubes reinforced bioplastic, *Macromol. Mater. Eng.*, **294**, 839-46 (2009).

[27]J.Y. Nam, S.S. Ray, and M. Okamoto, Crystallization behaviour and morphology of biodegradable polylactide/layered silicate nanocomposite, *Macromol.*, **36**, 7126–31 (2003).

[28]S. Iijima, Helical microtubules of graphitic carbon, *Nature.*, **354**, 56–58 (1991).

[29]J Gao, M. E Itkis, A Yu, E Bekyarova, B Zhao, and R Haddon, Continuous Spinning of a Single-Walled Carbon Nanotube–Nylon Composite Fiber, *J. Am. Chem. Soc.*, **127**, 3847-54 (2005).

[30]E. T. Michelson, C.B. Hoffman, A. G. Finzler, R. E. Smalley, R. H. Hauge, and J. L. Margrave, Fluorination of Single-Wall Carbon Nanotubes, *Chem. Phys. Lett.*, **296**(1-2), 188-94 (1998).

[31]D Qian, E. C Dickey, R Andrews, T Rantell, Load transfer and deformation mechanisms in carbon nanotube-polystyrene composites, *Appl. Phys. Lett.*, **76**(20), 2868-70 (2000).

[32]J. P Salvetat, A. J Kulik, J. M Bonard, G. A. D Briggs, T Stockli, K Metenier, et al, Elastic Modulus of Ordered and Disordered Multiwalled Carbon Nanotubes, *Adv. Mater.*, **11**, 161-65 (1999).

[33]C Park, Z Ounaies, K. A Watson, R. E Crooks, J Smith, and S. E Lowther, Dispersion of single wall carbon nanotubes by in situ polymerization under sonication, *Chem. Phys. Lett.*, **364**, 303-08 (2002).

[34]M. L. Chapelle, C. Stephan, T. P. Nguyen, S. Lefrant, C. Journet, P. Bernier, Raman characterization of single-walled carbon nanotubes and PMMA-nanotubes composites, *Synth. Mater.*, **103**, 2510-12 (1999).

[35]K Kim, S. J. Cho, S. T. Kim, I. J. Chin, H. J. Choi, Formation of Two-Dimensional Array of Multiwalled Carbon Nanotubes in Polystyrene/Poly(methyl methacrylate) Thin Film, *Macromolecules.*, 38, 10623-26 (2005).

[36]S. Sinha Ray, S. Vaudreil, A. Maazouz, and M. Bousmina, Dispersion of multi-walled carbon nanotubes in biodegradable poly(butylene succinate) matrix, *J. Nanosci. Nanotechnol.*, **6**, 2191–95 (2006).

[37]E.W. Fisher, H.J. Sterzel, and G. Wegner, Investigation of the structure of solution grown crystals of lactide copolymers by means of chemical reactions. *Kolloid Z.Z. Polym.*, **25**, 980-90 (1973).

[38]E.T. Michelson, C.B. Hoffman, A.G. Finzler, R.E. Smalley, R.H. Hauge, and J.L. Margrave, Fluorination of single-wall carbon nanotubes, *Chem. Phys. Lett.*, **296**, 188-94 (1998).

STRUCTURE AND PROPERTIES MULTILAYERED MICRO- AND NANOCOMPOSITE COATINGS OF Ti-N-Al/Ti-N/Al$_2$O$_3$

A.D. Pogrebnjak[1,2], V.M. Beresnev[6], M.V.Il'yashenko[1,2], D.A. Kolesnikov[3], A.P. Shypylenko[1,2], A.Sh. Kaverina[1,2], N.K. Erdybaeva[3], V.V. Kosyak[1], P.V. Zukovski[5], F.F. Komarov[4], V.V. Grudnitskii[6].

[1]. Sumy State University, Str. R-Korsakov, 2, Sumy 40007, Ukraine
[2]. Sumy Institute for Surface Modification PO BOX 163, 40030 Sumy, Ukraine
[3]. Belgorod State University, Belgorod, Russia
[4]. Belarus State University, Minsk, Belarus
[5]. Lublin University of Technology, Lublin, Poland
[6]. Kharkov National University Kharkov, Ukraine

ABSTRACT

This paper presents the first results on formation and studying of structure and properties of nanocomposite combined coatings. By means of the deposition processes modeling (deposition conditions, current density-discharge, plasma composition and density, voltage) we formed the three-layer nanocomposite coatings of Ti-Al-N/Ti-N/Al$_2$O$_3$/.The coating composition, structure and properties were studied using physical and nuclear-physical methods. The Rutherford proton and helium ion back scattering (RBS), Scanning Electron Microscopy with microanalysis (SEM with EDS and WDS), X-Ray diffraction (XRD) including a grazing incidence beam to 0.5°, as well as nanohardness tests (hardness) were used for analysis. Measurements of wear resistance, corrosion resistance in NaCl, HCl and H$_2$SO$_4$ solution were also performed. To test mechanical properties such characteristics of layered structures as hardness H, elastic modulus E, H^3/E^2 etc. were measured. It was demonstrated that the formed three-layer nanocomposite coatings had hardness of 32 to 36 GPa, elastic modulus of 328 ±18 to 364 ±14 GPa.

Its wear resistance (cylinder-surface friction) increased by a factor of 17 to 25 in comparison to the substrate (stainless steel). The layers thickness was in the range of 56 – 120 μm.

INTRODUCTION

Traditional methods of surface modification (which are: physical, chemical, electrochemical and mechanical ones [1]) as well as more advanced methods such as ion implantation, ion-assisted deposition of thin films, plasma technologies and electron beam treatment in some cases cannot result directly in a desirable way. In this connection for solving the industrial problems existing in ship building and chemistry, for instance [2], one has to combine such methods of surface modification for the production of hybrid coatings possessing the definite operation properties. An oxide-aluminum ceramics and other coatings based on titanium carbide and tungsten carbide and nitrides possess a number of useful properties, which are able to provide corrosion protection, high hardness and mechanical strength, low wear, and good electro-isolation properties [1-3].

It is well known that Ti-Al-N nano-composite coatings feature high physical and mechanical properties such as high hardness and elastic modulus. However, high hardness values are found only in coatings with small nano-grain sizes [4].

In ref. [4], we reported that deposition of Ti-Al-N coatings on to thick Ni-Cr-B-Si-Fe one resulted in improved physical-mechanical properties. In this case, hardness values reach only 22 ± 1.8GPa, which, first of all, was related to large nano-grain sizes of 17 to 22 and 34 to 90nm. A thin film with thickness less than 3.5μm was deposited on to Ni-Cr-B-Si-Fe thick coating of 60 to 70μm using magnetron sputtering with an alloyed Ti$_{40}$Al$_{60}$ target.

In the second paper [4,5], steel samples were coated with 2.5 μm coating by usage of vacuum-arc source in HF discharge. The fabricated coatings demonstrated high hardness reaching 35 ± 2.1 GPa combined with high wear resistance, scuffing resistance, and low friction coefficient in comparison with standard TiN.

EXPERIMENTAL

Samples of stainless steel 321 of $(2.5 \div 3)$ mm thickness were coated using plasma-detonation method by the device "Impule-6". The coating, with thickness of about 50μm, was fabricated from α-Al_2O_3 powder with 23 to 56μm grain size. Coatings were deposited within 20mm width, for one pass. Gas expenditures and battery capacity were similar to those applied in ref. [4]. After the surface purification by glow discharge, TiN coating of 1.8 to 2.2μm thickness was deposited on to Al_2O_3 coatings using 100 A arc current of Ti cathode and leaking-in N/Ar gas mixture.

Using the alloyed cathode TiAl, Ti-Al-N layers of different thickness in the range of $(2.2 \div 2.5)$ μm were deposited also in N/Ar medium. The resulting thickness of a three-layer multi-component coating reached 53 to 56.5 μm.

To analyze the coating structure, we used following methods: X-ray diffraction (XRD), TEM analysis, scanning electron microscopy with micro-analysis (SEM with EDS), and Rutherford back-scattering method (RBS) (He ions of 2.29 MeV and protons of 1.001 MeV) for composition analysis. Electron spectroscopy and corrosion tests were performed using a standard cell [4,5,6]. Wear resistance tests were performed according to the cylinder-plane scheme.

Transversal and angular cross-sections $(7 \div 10°$ angle) were prepared for several samples to analyze depth element distribution over the total multi-layered coating. They were used for electron microscopy, micro-analysis, point-by-point XRD-analysis, and nano-indentation.

To investigate the homogeneity of powder coatings and identify the various inclusions of the matrix substrate in the contact area for 10 minutes. The surface was etching using solution of hydrofluoric acid (50 ml HF, 50 ml H_2O). The structure of the steel studied in the transition region was determined after subsequent grinding and etching (t = 10 min.) Nitric acid solution (2 mL HNO_3, 48 mL ethyl alcohol).

The elemental distribution in the coatings, before and after corrosion treatment, was evaluated by the Rutherford Back-scattering Spectroscopy (RBS) at the 5.5 MVTandemAccelerator of the NCSR DEMOKRITOS/Athens using a deuteron beam of 1.5 MeV energy (scattering angle: 170°, solid angle: 2.54×10^{n3} sr). Nuclear reaction analysis using the same accelerator was used for the determination of the N depth distribution on the samples ($^{14}N(d, a)^{12}C$ nuclear reaction, Ed =1.350 MeV, detector angle: 150°, detector solid angle: $2.54 \times 10''3$ sr) .

For the RBS and NRA measurements a C. Evans & Assoc. scattering chamber equipped with a computer controlled precision goniometer and a laser positioning system was utilized. The vacuum in the scattering chamber was constant (ca. $2 \times 10^{\sim7}$ Torr). The beam current on the target did not exceed 10 nA, while the beam spot size was 1.5 mm×1.5 mm.

The analysis of the NRA and RBS data was performed using the simulation code RUMP [7].

The corrosion resistance of the prepared coatings was investigated using electrochemical techniques. An AUTOLAB Potentio-Galvanostat (ECO CHEMIE, Netherlands) and Princeton Applied Research corrosion testing cell were used for the electrochemical measurements. A saturated calomel electrode used as a reference electrode and a graphite one as an auxiliary electrode for all measurements.

The tests in 0.5 M H_2SO_4 solution were carried out in the potential region -1000 to $+1500$ mV at ambient temperature. Five rapid scans (scan rate=25 mV/s) followed by one slow scan (scan rate=0.25 mV/s) were performed on each specimen. The rapid scans allow investigations under constant conditions of the material surface and corroding medium, whereas slow scans lead to predictions of the general corrosion behavior of the material. In all cases the sample surface exposed to the corroding medium was 1 cm^2 [5-7]. The above-mentioned experimental conditions were also applied for the corrosion tests in HCl and NaCl solutions, whereas the scanning region was from -300 to $+1700$ and from -1000 to $+1000$ mV, respectively [5].

RESULTS AND DISCUSSION

Table 1 presents the results of calculation for nano-hardness and elastic modulus for every layer of the multi-layered system. One can see that nano-structured Ti-Al-N featured the highest hardness of 35 ± 1.8 GPa and the highest elastic modulus 327 ± 17 GPa. Evaluations of grain sizes performed according to Debay-Sherer method, demonstrated that grain size in near surface layer was in the range of 10 to 12 nm, at the same time, the second TiN layer demonstrated sizes of 30 to 35nm and Al$_2$O$_3$ coating demonstrated wide variation of values from units of a micron (5% of grains) to tens of micron (25%) and grains exceeding 100nm (less than 20%). Thus, the third layer of Al$_2$O$_3$ ceramics turned out to be dispersion-hardened rather than nano-structured one.

Table 1. The values of hardness and modulus of elasticity, the size of layers sandwich of nanocomposites combined coatings deposited on stainless steel

Composition of coating (layer)	H, GPa	E, GPa	Grain size (nm)	Thickness of layers (µm)
Ti-Al-N	35±1,8	327±13	10–12	2,2–2,5
Ti-N	22±6	240±16	20–35	1,8±0,2
Al$_2$O$_3$	16 – 20	194±8	10^4–10^5	48–52
Steel	4,2	46±2,5	10^5–10^6	$2 \cdot 10^3$

Figure 1 shows an image of a transversal cross-section of a thin Ti-Al-N coating and that of a thick one Ni-Cr-B-Si-Fe . An EDS microanalysis was performed over the cut section, and the results are presented in Fig.2a, b, c. The presented spectra indicate that in the thin coating only titanium and aluminum were present.. The interface of the thin film and the coating contained Ti, Al, N,Ni, Cr, Fe, and Si at some places. In the thick coating we found Ni, Cr, Fe, Si, Ni amounting 45%, and all the rest was the mentioned above elements.

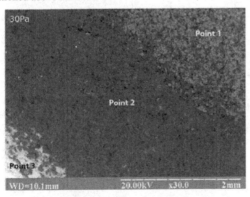

Fig.1 The image of coating cross-section prepared at an angle 7-10° multilayer nanocomposite coatings based on Ti-Al-N/Ni-Cr-B-Si-Fe.

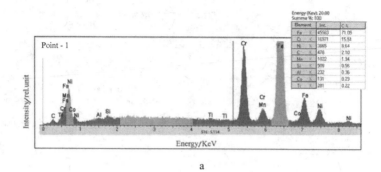

a

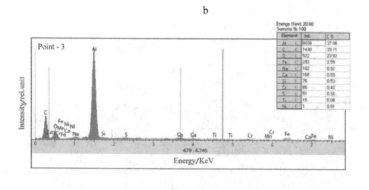

b

c

Fig.2 – The Energy-dispersive spectra obtained from the "angle lap" sections of multilayer surface beginning from the interfacial area of the Al_2O_3 coating (substrate (a), the second layer (b), third (the top layer (a)) shown in Figure 1.

Figure 3 shows the RBS spectra of He ions (a) and protons (b). In this coating, Ni, O, Al, Ti elements and small concentration of Nb atoms, as well as very small amount of Ta were observed. The latter came from chamber walls as non-controlled impurity. When the spectrum formed a step, the compound stiochiometry, which was obtained from RBS spectra according to formulas [6] demonstrated $Ti_{60}Al_{40}$ values. The used proton beam energy was sufficient to analyze the second layer of TiN over its depth and the next one of Al_2O_3 including the interface of Al_2O_3 layer (relative to the substrate). The analyzing proton beam could not reach the back Al_2O_3 interface because of insufficient energy. The images of coating cross-sections showed three layers of different thickness.

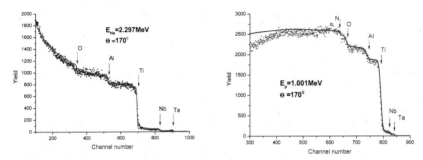

Fig.3. RBS experimental spectra obtained for multi-layered Ti-Al-N/Ti-N/Al_2O_3 coating. Helium ion energy was 2.297MeV (a) and proton energy was 1.01MeV (b).

Visual analysis indicated good quality of the obtained coatings, which did not include pores but exhibited surface roughness resulting from the plasma-detonation treatment. All subsequent layers repeated this roughness, diminishing it a little degree due to smoothed interfaces between protuberances and valleys. However, after deposition of the thick Al_2O_3 we polished surface was carried out on the grinding machine with diamond paste. for RBS measurements. The cross-sections were prepared for micro-analysis and nano-indentation tests. Point-by-point micro-analysis, which was performed in angular cross-section starting from the substrate, demonstrated that the first layer was composed of Ti-Al-N, the second one – of TiN, and the third one of Al_2O_3.

Figure 4 shows diffraction patterns for multi-layered nano-composite coatings of Ti-Al-N/Ti-N/Al_2O_3 at the initial state. As it is seen in this Figure, Al_2O_3, TiN, $AlTi_3N$, (AlTi)N phases are presented in this coating. $Cr_{0.19}Fe_{0.7}Ni_{0.11}$ phase, which came from the substrate, was also possible.

After 600°C annealing, the coating phase composition was changed. However, 900°C annealing during 3 hours in air (the diffraction pattern in Fig.4, the upper curve) resulted in formation of TiO_2 and phase of Al_2O_3 became more micro-crystalline and contained only α-Al_2O_3, i.e. total oxidation of Ti and Al occurred as a result of 900°C annealing in air. Coating hardness also decreased from 14.8 GPa to 8.8 – 12 GPa. In such a way, the upper two layers were oxidized (probably the first one totally and the second one partially). The layer, which was composed of Al_2O_3, did not demonstrate transition from γ-phase to α-one, possibly because it starts at temperatures higher than 900°C.

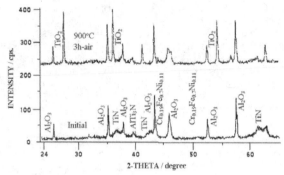

Fig.4. Diffraction patterns for multi-layered nano-composite Al-N/Ti-N/Al₂O₃ coating after deposition (in several weeks) and after 900°C annealing in air during 3 hours.

Auger-electron microscopy, which was performed for these coatings, demonstrated Ti, N, Al, O, and C elements. At the initial moment of deposition, the concentration of these elements was changed, but after the etching during more than 12 min, their profile did not change. Additional studies performed using nuclear reactions method, allowed us to obtain depth element profiles for two upper layers including the interface between the second and the third layer. Thickness of Ti-Al-N and Ti-N layers was accurately measured. It amounts to 4.27μm. Measurements after 900°C annealing of the multi-layered system demonstrated essential changes of element concentration profiles over depth. The upper layer was enriched with oxygen and carbon, that indicated formation of Ti and Al oxides. This was confirmed by XRD analysis.

Fig.5. (a) and (b) show images for coating surfaces before and after annealing in air at 600°C, and for etched angular cross-sections. These images demonstrated no essential changes in the coating structure and element composition.

Fig.5. Structure of transversal cross-section in multi-layered nano-micro-composite coating (left) and surface image (right) in the initial state after deposition (in three months).

Measurements of lattice parameters for TiN, (Ti, Al)N, and Al₂O₃ coatings demonstrated that 600°C annealing decreased stresses arising in lattices: both macro-stresses at film-film and film-thick coating interfaces and micro-stresses arising in nano- and micro-grain structures.

Figure 6 a, shows the concentration profiles of the depth distribution of titanium atoms in hybrid coatings. According to the results the minimum concentration of titanium atoms in the surface region has coatings in the initial state. The maximum concentration of titanium atoms (about 52 at.%) was observed at the distance of 1.4 μm from the surface.

High concentrations of oxygen and aluminum were found in the subsurface area of titanium nitride. Depending on the mode of electron-beam melting of metal-ceramics underlayer the percentage and the depth distribution of elements changes (Fig. 6 b, c, d). For example, in the surface of the hybrid coatings after electron-beam exposure with the power density of 360 W/cm^2 in plasma-detonation sublayer of aluminum oxide in the titanium nitride film, aluminum atoms are missing.

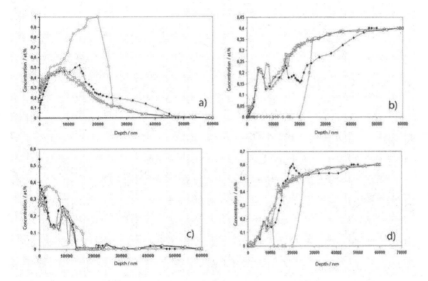

Fig.6 - The depth distribution of the constituent elements of the hybrid coatings surface: a)– Ti; b – N; c) – Al; d) – O ($-\!\!\bullet\!\!-$ without electron-beam melting; $-\Box-$ - 240 W/cm^2; $-\triangle-$ - 300 W/cm^2; $-\diamond-$ - 360 W/cm^2).

Figure 7 shows results of wear tests, which were performed according to the plane-cylinder scheme. These results demonstrated that the most essential wear occurred in the substrate surface (curve 1). After deposition of Al$_2$O$_3$ coating by plasma-detonation technology (curve 2) wear sharply decreased. In the case of TiN deposition, wear was lower than in the case of Al$_2$O$_3$ (curve 3). The lowest wear was found in the case of multi-layered coating Ti-Al-N/Ti-N/Al$_2$O$_3$ (curve 4) [5-7].

Corrosion tests performed in Electro-Chemical Lab (Thessaloniki, Greece) using International standards in 0.5 M H$_2$SO$_4$ solution and by simple micro-weighing after definite time period (3 to 6 months) in NaCl and HCl solutions, demonstrated high coating resistance in comparison with that of stainless steel 321 substrate (European standards).

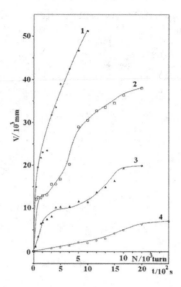

Fig.7. Dependences of material wear tested by cylinder friction over the sample surfaces: 1 – initial state; 2 – Al$_2$O$_3$ coating; 3 – TiN/Al$_2$O$_3$ coating; 4 – multi-layered nano-micro-composite coating of Al-N/Ti-N/Al$_2$O$_3$.

CONCLUSION

The fabricated multi-layered nano-micro-composite coatings based on Ti-Al-N-Ti-N/Al$_2$O$_3$ system featured thermal stability in air till 900° C. They also featured high wear resistance under cylinder friction over their surfaces and high corrosion resistance in NaCl and H$_2$SO$_4$ medium. However, 900° C annealing resulted in total oxidation of the upper Ti-Al-N layer and partial oxidation of Ti-N one. The coating hardness decreased more than by factor of 2. At the same time, electron pulsed beam (without melting) did not decrease surface hardness, possibly, due to its short-time action. However, it results in redistribution of impurities (coating components) at interfaces of this multi-layered coating.

ACKNOWLEDGEMENTS
The work was fulfilled within the frameworks of ISTCs K-1198 Project and partially within the project for NAS of Ukraine "Nano-Systems, Nano-Composites, and Nano-Materials." The authors are thankful to colleagues Yu.A.Kravchenko, A.D.Mikhaliov, and V.S.Kshnyakin from Sumy Institute for Surface Modification, Dr.S. Kislitsyn from Institute of Nuclear Physics of National Nuclear Center of Kazakhstan for their help in experiments, as well as Dr.Noly, Prof. Misaelides from Thessaloniki, Greece for their corrosion tests.

REFERENCES

[1]J. Musil, H. Hruby, Superhard nanocomposite $Ti_{1-x}Al_xN$ films prepared by magnetron, *Thin* Solid *Films*, **365**, 104-109 (2000)

[2]Zhu J. Ye L. Li Wang F., Fabrication of TiN/Al_2O_3 Composites by a Novel Method, *Appl. Mech. and Mater*. **44-47**, 2504-2508 (2011)

[3]P. Budzynski, J. Sielanko, Z. Surowiec and P. Tarkowski, Properties of (Ti,Cr)N and (Al,Cr)N thin films prepared by ion beam assisted deposition, *Vacuum*. **83**. S186-S189 (2009)

[4]A.D.Pogrebnjak, A.A.Drobyshevskaia, V.M.Beresnev, M.K.Kylyshkanov, G.V.Kirik, S.N.Dub, F.F.Komarov, A.P.Shypylenko, Yu.Zh.Tuleushev, *Jour.Tech.Phys*, **81**, (2011).

[5]V.M. Beresenev, A.D. Pogrebnyak, P.V. Turbin, S.N. Dub, G.V. Kirik, M.K. Kylyshkanov, O.M. Shvets, V.I. Gritsenko and A.P. Shipilenko, Tribotechnical and Mechanical Properties of Ti-Al-N Nanocomposite Coatings Deposited by the Ion-Plasma Method, *Journal of Friction and Wear*, **31**, №5, 349-355 (2010).

[6]A.D. Pogrebnjak, Yu.A. Kravchenko, S.B. Kislitsyn, Sh.M. Ruzimov, F. Nolid, P. Misaelides and A. Hatzidimitriou, $TiN/Cr/Al_2O_3$ and TiN/Al_2O_3 hybrid coatings structure features and properties resulting from combined treatment, *Surface and Coatings Technology*, **7**, 2621-2632 (2006).

[7]A.D. Pogrebnjak, M.M. Danilenok, A.A. Drobyshevskaia, V.M. Beresnev and N.K. Erdybaeva, et.al. Investigation of the structure and physicochemical properties of combined nanocomposite coatings based on Ti–N–Cr/Ni–Cr–B–Si–Fe, *Russian Physics Journal*, **52**, 1317 – 1324 (2009)

[8]Pogrebnjak A.D., Shpak A.P., Azarenkov N.A., Beresnev V.M. Structures and Properties of Hard and Superhard Nanocomposite Coatings//Usp. Phys. **179**(1), 35-64. (2009)

FORMATION OF NANOSTRUCTURED CARBONITRIDE LAYERS DURING IMPLANTATION OF NiTi

Alexander Pogrebnjak
Sumy State University, Sumy, Ukraine

Sergey Bratushka
Sumy Institute for Surface Modification, Sumy, Ukraine

Nela Levintant
Institute of Fundamental Technological Research, Warsaw, Poland

ABSTRACT

The surface layer of an equiatomic TiNi alloy, which exhibits the shape memory effect in the martensitic slate, is modified with high-dose implantation of 65-keV N^+ ions (the implantation dose is varied from 10^{17} to 10^{18} ions/cm^2). TiNi samples are implanted by N^+, Ni^+-N^+, and Mo^+-W^+ ions at a dose of 10^{17}-10^{18}cm^{-2} and studied by Rutherford back scattering, scanning electron microscopy, energy dispersive spectroscopy, X-ray diffraction (glancing geometry), and by measuring the nanohardness and the elastic modulus. A Ni^+ concentration peak is detected between two maxima in the depth profile of the N^+ ion concentration. X-ray diffraction (glancing geometry) of TiNi samples implanted by Ni^+ and N^+ ions shows the formation of the TiNi (B_2), TiN, and Ni_3N phases. In the initial state, the elastic modulus of the samples is E = 56GPa at a hardness of H = 2.13 ± 0.3 CPa (at a depth of 150 nm). After double implantation by Ni^+-N^+ and Mo^+-W^+ ions, the hardness of the TiNi samples is 2.78 ± 0.95 GPa at a depth of 150 nm and 4.95 ± 2.25 CPa at a depth of 50 nm; the elastic modulus is 59 GPa. Annealing of the samples at 550°C leads to an increase in the hardness to 4.44 + 1.45 GPa and a sharp increase in the elastic modulus to 236 ± 39 GPa. A correlation between the elemental composition, microstructure, shape memory effect, and mechanical properties of the near-surface layer in TiNi is found.

INTRODUCTION

What makes na no structured films unique is the high volume fraction and strength of interfaces, the absence of dislocations inside the crystallites, the possibility of changing the ratio of the volume fractions of the crystalline and amorphous phases and the mutual solubility of metal and non-metal components. The presence of a large interfacial area (the volume fraction may reach 50%) in nanostructured films allows one to change their properties either by material and electronic structure modification or by doping. The strength of the interfaces favours the stability of nanostructured films against deformation. The lack of dislocations inside the crystallites increases the elasticity of these films. These factors altogether allow the production of nanomaterials with improved physicochemical and physicomechanical characteristics, namely, with high hardness $(H > 30$ GPa), elastic recovery $(W_e > 70\%)$. strength, thermal stability, and heal and corrosion resistance.

An important feature of the ultrahard nanomaterials based on muliicomponent (multifunctional) nanostructured films (MNF) is that materials with the same hardness may differ in the elastic modulus (E), resistance against elastic strain to failure (H/E) and resistance against plastic deformation $(H^3/E^2)^{1-3}$.

Multifunctional nanostructured films are used to protect surfaces of articles and tools exposed simultaneously to elevated temperatures, corrosive media and various types of wear. This refers, first of all. to the culling and press forming tools, mill rolls, parts of aircraft engines, gas turbines and compressors, friction bearings, glass and mineral fibre extrusion nozzles, *etc.* These films are also indispensable in the development of new-generation biocompatible materials, namely, orthopedic and dental implants, craniofacial and maxillofacial surgery implants, fixation of cervical and lumbar spines and so on.

Currently, titanium nitride-based films are widely used in industry. The introduction of a third component (for example - Si or B) into the films may markedly improve their physico-mechanical properties and thus extend the scope of application. The interest in the Ti-Si-N films is mainly caused by their high hardness,[2] thermal stability[4], resistance against high-temperature oxidation[5] and substantial abrasive wear resistance. Nanostructured Ti - B - N films possess a series of important performance characteristics including high hardness, thermal stability up to 100^0C (in vacuo), heat, wear and corrosion resistance, impact resistance and high electrical resistance[6-8]. Chromium is known to have a beneficial effect on the stability of titanium carbides, borides and nitrides against oxidation and on the wear resistance of articles made of them at elevated temperatures.

The application of high-dose intense implantation leads to an increase in the ion (mainly N^+ ion) penetration depth; intensified scattering of the surface layer; a shift in the maximum, concentration, and shape of the concentration profile; and many other processes that are weakly pronounced during low-intensity ion implantation at low doses (several units of atomic processes) of implanted ions[8-10]. On the other hand, TiNi-based alloys belong to the group of materials in which a high-temperature phase with a B_2 structure undergoes a shear or martensitic phase transformation as the temperature changes or a stress is applied. The atomic restructuring in TiNi-based alloys is accompanied by both martensitic anelasticity effects and a change in their surface state, which is caused by the complex structure of the martensite phase in them[10, 12].

As a result, a developed martensitic relief with a large number of various interfaces appears, which should affect both the electrochemical and corrosion properties and the plasticity and strength properties of these materials. As a method of surface alloying, ion implantation of a surface can strongly affect the structural parameters and stability of the B_2 phase in the near-surface layers and, hence, the following set of its properties: the martensite transformation temperature, the martensite anelasticity parameters, the shape memory effect (SME), and superplaslicity. As a result, it can change the deformation relief, the cracking conditions, and the electrochemical and corrosion properties[9]. Therefore, double implantation of N^+ and Ni^+ ions into TiNi is of particular interest, since the implantation of Ni^+ ions changes the equiatomic composition of the alloy and. in combination with N^+ ions, hardens the surface layer and. correspondingly, modifies the physicomechanical and chemical properties[13]. The effect of N^+ ions on the mechanical properties of steels and alloys is well known. The alloying of steels and alloys with elements such as W and Mo is widely used to improve their mechanical properties; therefore, it is interesting to perform implantation of Mo^+ and W^+ ions at high doses. The purpose of this work is to study the depth profile of the elemental composition of the implanted layer, the structure and morphology of the TiNi alloy surface, and the mechanical properties of the alloy implanted by high doses of N^+ ions (10^{17}—10^{18} cm^{-2}) and subjected to double implantation of N^+, $Ni^+ + N^+$ and $W^+ + Mo^+$ ions at doses ranging from 5×10^{17} to 10^{18} cm^{-2}.

METHODS OF ANALYSIS AND COATING APPLICATION

We analyzed equiatomic TiNi (51.5% Ni) alloy samples 22x5.4x0.25 m in size. In the initial state, the samples were vacuum annealed at 803 K for 30 min followed by slow cooling. After cooling, the sample surfaces were etched with a mixture of 10% $HClO_4$ and 90% acetic acid. The N^+ ion implantation of the TiNi samples was performed on a semi-industrial IMJON (Warsaw) implanter at doses of 1×10^{17}, 5×10^{17}, and 10^{18} cm^{-2} at a current density of 0.8-1 mA. The implantation of Ni^+ and $Mo^+ + W^+$ ions was carried out using a vacuum-arc Diana source at a voltage of about 60 kV, a dose of 5×10^{17} cm^{-2}, and a substrate temperature of less than 250°C. Irradiation was performed at a pressure of $\sim 10^{-3}$ Pa. The pulse duration was 200 μs, the pulse repetition frequency was 50 Hz, and the nitrogen concentration in TiNi was determined from the "eating away" in its energy spectrum. The phase-transformation temperatures were determined with a Pyris-1 differential scanning calorimeter, and the elemental composition of the samples was determined by the following methods: Auger electron spectroscopy on a PHI-660 (Perkin-Elmer) device, scanning electron microscopy on a Selmi (Sumy. Ukraine) microscope equipped with EDS and WDS microanalyses and on a Perkin-Elmer microscope, and Rutherford backscattering of ions (2.012-MeV proton beams. 2.035-MeV $^+$He ion beams). Rutherford backscattering spectra were analyzed using the standard RUMP and DWBS software packages[11] to construct the depth profiles of elements (Ni^+, Mo^+, W^+ ions). Ions backscattered at an angle of 170° were detected using a surface barrier detector with an energy resolution of 20 keV. In addition, we used a Neophot-2 optical microscope. To measure the mechanical properties and SME, we used a diamond pyramid (Brinel) microindenter) with a side of 40 nm at a load of 4, 7, 10, 13, 16 and 20 N and a Talysurt-5-120 scanning profilometer. The measurements were performed in the initial state and after ion implantation.

The microhardness on the surface of a sample and across it was measured with a PMT-3 device at various loads. The nanohardness was measured with a trihedral Berkovich pyramid on a Nano Indenter II (MTS System Corp., Ridge. Tennessee, United States) nanohardness tester. To find the hardness and the elastic modulus at the maximum load, we used the Oliver-Pharr technique.

X-ray diffractometry, using the Philips diffractometer type X'Pert in the Bragg-Brentano geometry, was used to identify the phase composition of NiTi alloy samples in both unimplanted conditions and after implantation with nitrogen. CuK radiation (wavelength λ - 0.154184 nm) diffracted by the sample was selected by a graphite monochromator. The scanning voltage of the X-ray tube was 40 kV, the current was 25 mA, the exposure time was 10 s and the measured angle, 2Θ, was from 25° to 95°. The scanning step was 0.02°. The low temperature X-ray diffraction studies were carried out using the TTK Low-Temperature Camera (Anton Paar). The sample was heated from -50°C up to +150°C in argon atmosphere. The measured angle, 2Θ, was from 35° to 47° with the scanning step of 0.02°. Lattice parameters were determined by using the Philips X'Pert Plus software for all detected peaks.

Specimens for TEM and HREM (Tecnai G^2. FEI Company) examinations were prepared by Focused Ion Beam System (FEI QUANTA 3D).

EXPERIMENTAL RESULTS AND DISCUSSION
The studies were started with DSC measurements. The sample weight of approximately 5 mg was analyzed with an empty aluminum pan as the reference. A temperature range from -50 °C to +150 °C was scanned at a rate of 20 °C/min during cooling and heating. Fig. 1 shows the evolution of DSC cooling/heating curves for the virgin and ion-implanted alloys. As it can be seen, the NiTi alloy transforms in two steps showing two peaks on the DSC curve in the cooling direction. The first DSC peak correlates with the transformation from the austenite (A) with a cubic structure (B_2-phase) to the R-phase with a rhombohedral one (referred to as a rhombo-hedral distortion of the austenite). The second DSC peak correlates with the transformation from the R-phase to the martensite (M) with a monoclinic structure (B_{19}-phase). Similar two-step martensite phase transition from high temperature was observed for the NiTi alloys after the thermo-mechanical treatment or solution treatment and subsequent aging. The one-step but composed transition took place during the heating process. The endothermic peak during the heating process resulted in the transition of the martensite to the austenite phase. The peak in the heating direction corresponds to the austenite (B_2-phase) with t_{start} = 51.6°C, t_{finish} = 62.3 °C for the virgin alloy and with t_{start} = 52.5 °C, t_{finish} = 64°C for the ion-implanted one. The subscript 'start' denotes the onset temperature at which the phase transformation starts, and the 'finish' the temperature at which the phase transformation finishes.

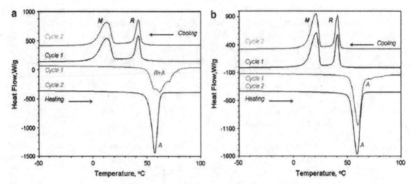

Figure 1. DSC curves of the (a) virgin and (b) ion-implanted alloys indicating a two-step phase transformation occurred during cooling.

The two peaks in the cooling direction corresponding to the R-phase and the martensite $B_{19'}$-phase were determined to have transformation temperatures, respectively: t_{start} = 45.1°C, t_{finish} = 37.8 °C, t_{start} = 19 °C, t_{finish} = 1°C for the virgin alloy and t_{start} = 43.3 °C, t_{finish} = 37.4°C, t_{start} = 24.6 °C, t_{finish} = 11.6 °C for an ion-implanted alloy. With Auger electron spectroscopy, we also studied the TiNi samples before and after implantation with nitrogen ions (Fig. 2). It is seen that, in the initial state, carbon and oxygen are present near the surface; after sputtering for 15-18 min. only nickel and titanium are present in the NiTi sample and their concentrations are close to the equiatomic composition. After implantation, the nickel concentration in the surface layer decreases to almost 10 at % because of sputtering of the surface. Since it is difficult to separate the peaks of nitrogen and titanium ions, we constructed a TiN profile with Auger electron spectroscopy.

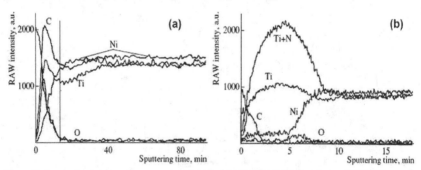

Figure 2. Auger electron spectroscopy data for (a) unimplanted TiNi samples and (b) TiNi samples implanted at an energy of 65 keV and a dose of 5×10^{17} ion/cm^2.

This profile indicates that the penetration depth of N$^+$ ions is about 280-200 nm. After sputtering for 10 min the concentrations of nickel and titanium ions are seen to level off and reach their intrinsic levels (49.9 and 50.1 at. % Ni and Ti in the crystal lattice of the TiNi alloy, respectively).

Figure 3 shows optical microscopy data. They demonstrate a typical martensitic structure and changes in the martensitic structure after implantation.

Figure 3. Optical micrographs of (he cross section of a TiNi alloy with a martensitic structure and SME: (a) before and (b) after ion implantation.

The TiNi samples implanted by N^+ ions are seen to have a higher (by 15-20%) hardness than the initial samples. The changes in the SME (Fig. 4) demonstrate that the implanted TiNi alloy exhibits an indentation with a higher hardness after recovery as a result of healing to 75°C. In other words, all mechanical changes related to the SME and mechanical properties (hardness) are interrelated with the elemental composition and micro-structure of the material.

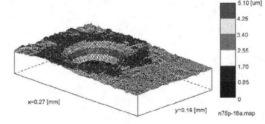

Figure 4. SME measurements (with a scanning profilomeier) on a TiNi sample after heating to 75°C.

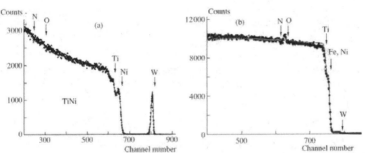

Figure 5. (a) Rutherford backscattering spectrum (2.035-MeV helium ions) of a TiNi sample after 60-keV N ion implantation at a dose of 10^{18} cm^{-2} followed by implantation of Ni ions at a dose of 5 x 10^{17} cm^{-2} and a voltage of 60 kV. (b) The spectrum recorded on the same sample with protons at an energy of $E = 2.012$ MeV.

Figure 5a shows the spectrum of helium ions sputtered by the implanted sample. Among the light elements, only oxygen exhibits a peak; however, oxygen is located near the sample surface and the eating away is caused by nitrogen. Figure 5b shows the spectrum of protons scattered by the initial implanted sample. Although titanium is seen to be distinguished from nickel and iron, light impurities are not detected. The surface layer of the implanted sample has a significant concentration of nitrogen and oxygen and the characteristic eating away is observed in the spectrum of TiNi implanted sequentially by nitrogen and nickel. We used a standard computer program and determined the nitrogen concentration from the eating away in the spectrum and plotted element-concentration profiles. The eating away appears when a light element is added to a heavy-element matrix. The yield in a layer containing a light element decreases in proportion to a decrease in the heavy-element concentration. If the light-element concentration increases to 100%, a pure light-element layer forms. Therefore, we can easily determine the error in determining the concentration of nitrogen ions; it was found to be 5 at.%. The error in determining the nickel concentration for concentration profiles was 1.2 at %. The nickel profile was plotted beginning from 50 at %. This point was taken as an initial point, since the titanium or nickel concentration in the initial state is about 50 at.%. When analyzing the profile of N^+ ions in TiNi (Fig. 6), we see that the nitrogen profile has a double-humped shape: one concentration maximum is located near the surface (the maximum concentration is about 36 at. %). and the second peak is located at a depth of more than 130-150nm (133 at Å^2 and has a lower concentration (27 at. %). In the valley between the two nitrogen concentration maxima, the concentration of Ni^+ ions is maximal (about 20 at. %).

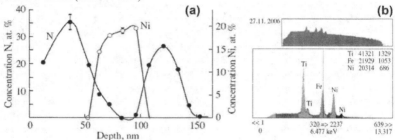

Figure 6. (a) Depth profiles of the element concentrations obtained from energy spectra of the TiNi sample surface after double implantation of N ions at a dose of 10^{18} cm^{-2} and Ni ions at a dose of 5x10^{17} cm^{-2} and (b) X-ray spectrum taken with WDS from the TiNi sample after double implantation by N and Ni ions.

Moreover, preliminary SIMS (secondary-ion mass spectrometry) studies also demonstrate the formation of a double-humped nitrogen concentration profile. However, the nitrogen concentrations in the maxima differ slightly from the RBS data, which is caused by the higher threshold of delegability of SIMS ($\sim10^{-5}$ at. %).

The X-ray diffraction patterns of the ion-implanted NiTi alloy against the temperature are shown in Fig. 7. From the results obtained it can be seen that alloys at 20 °C exhibit the three phases: the dominating $B_{19'}$-phase and a small amount of the R_2 and B_2 phases. The appearance of austenite phase (crystalline, nanocrystalline and/or amorphous-like) may be related to structural changes in the NiTi alloy during the ion implantation process and to high temperature of target. Appearance of the R-phase, similar to the results obtained for annealed materials, may be induced by high temperature of target. An increase of temperature in the material results in an increasing fraction of the B_2-phase. As follows from Fig. 7 the diffraction patterns of this $B_{19'}$-phase do not vary up to 65°C.

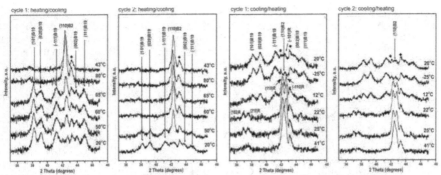

Figure 7. Recorded X-ray diffraction patterns versus temperature of ion-implanted NiTi alloy at a fluence of 10^{18} cm^{-2} and energy 50 keV.

Above this temperature an increasing fraction of the B_2-phase and a distinctly decreasing fraction of the $B_{19'}$ one were observed. The $B_{19'} \rightarrow B_2$ phase transformation finishes above a temperature of 80 °C. The X-ray diffraction patterns obtained for the highest temperatures (from 80°C to 150°C) contain only reflections from the B_2-phase.

Lowering the temperature up to about 41 °C, besides still existing B_2-phase, the R-phase appears. From the results obtained it can be seen that the R $\rightarrow B_{19'}$-phase transformation starts form a temperature of 25 °C and finishes about 25°C. Along with the decreasing temperature, besides the dominating $B_{19'}$-phase, a small amount of the B_2-phase was detected. Such phase composition was also observed for the material at 20 °C on the start and finish of the second thermal cycle. The broad structure of this diffraction peak may testify that still some amount of the amorphised and/or nano-crystalline B_2-phase is present in the alloy. It should be noted that from the results obtained there is no evidence for TiN and Ti_2Ni compound formations.

The concentration and the type of defects produced in materials during ion implantation depend on the implantation conditions, such as implantation temperature, fluence and fluence rate. Physical processes, which arise during interaction of accelerated ions with a crystal as a result of ion implantation, apart from the ionization of parent atoms and insertion of foreign ones, include formation of post-implanted crystal defects. It is very well known that when a material is uniformly irradiated by ions, isolated damaged regions are created in the first stage. Depending on the size of these regions and the irradiation fluence, the damaged regions start to overlap and a continuous disordered (saturated at the critical fluence) layer is formed. The disordered material contains large concentration of vacancy clusters, implanted impurity and other defects mainly within this material, but they are also present in transition zones. Naturally, the maximum of post-implanted defects distribution is closer to the crystal surface than the maximum of impurity concentration profile.

In the process of nitrogen ion implantation with the fluence 10^{18} cm^{-2} and 50 keV of energy, the well-defined double-layer structure with different microstructure as well as different phase and chemical composition was formed (Fig. 8). During ion implantation, collisions between the incident ions and the NiTi-target atoms lead to the formation of near-surface amorphized layer (A-layer, Figs. 8 and 9a) and the extended defects in the crystalline structure of bulk material (D-layer in Figs. 8a and 9b; Bulk in Fig. 8a).

The transition zone of a damaged region is wide and its composition changes gradually from totally amorphized Ti-rich material (heavily damaged and nano-crystalline) to Ni-rich crystalline (A1 and A2 regions in A-layer in Fig. 8b, c). Amorphous-like layer contains some amount of crystalline inclusions within its bulk (P1 and P2 regions in Fig. 9a), mainly near a bottom boundary of the transition zone (P2-region). This confirms the fact that the amorphization process occurs faster from the depth of maximum damage regions towards the sample surface than in its bulk. Due to differential strain between the undamaged and damaged layers, many cracks appear in this region. In the depth of 80-160 nm the material has a defected crystalline microstructure (Fig. 9b).

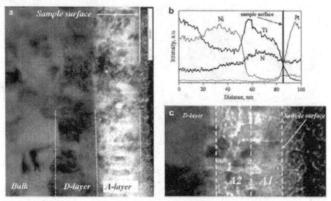

Figure 8. (a) Bright-field image of the ion-implanted NiTi alloy demonstrating the structure and (b), (c) chemical composition changes in near-surface layers.

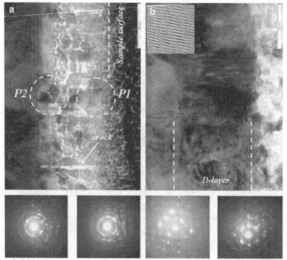

Figure 9. Bright-field image of the ion-implanted NiTi alloy demonstrating the details of structural changes in near-surface layers.

The hardness (H) and Young's modulus *(E)* data change correspondingly. Their values were determined at various depths (see Tables I, II). The hardness is seen to increase from 2.13±0.13 to 2.78 ± 0.95 GPa at a depth of 150 nm. The elastic modulus varies from E = 56±2 to 59±11 GPa. After annealing at 550°C for 2 h, these values increase sharply: H= 4.11± 0.35 GPa, E = 289 ± 81 GPa. Small differences between the hardnesses and elastic moduli of the initial and implanted and annealed samples are also detected at a depth of 50 nm. Upon implantation, we have H=4.96 ± 2.26 and E = 59±8 GPa; after annealing, we have H = 4.44±1.45 and E = 236±39 GPa (the elastic modulus increases by a factor of almost 4.5).

Table I. Hardness and elastic modulus of TiNi samples at a depth of 150 nm

	Sample	E, GPa	H, GPa
1	Initial	56 ±2	2.13 + 0.30
2	W + Mo	59 + 11	2.78 + 0.95
3	W + Mo. after annealing	298 + 81	4.11 ± 0.35

Table II. Hardness and elastic modulus of TiNi samples at a depth of 50 nm

	Sample	E, GPa	H, GPa
1	Initial	56 ±4	2.74 + 0.30
2	W + Mo	59 + 8	4.95 + 2.26
3	W + Mo. after annealing	236 + 39	4.44 ± 1.45

Figure 10 shows the TiNi sample surface after double implantation by N^+ and Ni^+ ions. The surface is rather rough *(a* = 0.8-1.2 μm) due to sputtering mainly by nitrogen atoms. This surface was subjected to electron- probe microanalysis. The following elements were detected in the near-surface layer (Fig. 10): N (~2.1 at. %), O (~5.61 at. %, C (~0.58 at. %), Ni (~49.43 at. %), and Ti (~41 at. %). To detect Ti, we used another detector.

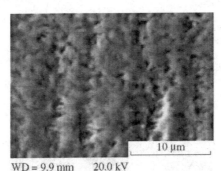

WD = 9.9 mm   20.0 kV

Figure 10. TiNi sample surface implanted by nitrogen and nickel.

The low (as compared to the RBS data) titanium concentration is caused by the larger depth of SEM as compared to EDS (which is 2.2 μm); moreover, the nitrogen ion range is at most 300 nm at the energies (60-70 keV) used in our experiments. Upon implantation, the elastic modulus increases insignificantly; however, after heat treatment, it increases to 236 ± 39 or 289 ± 80 GPa (i.e., by a factor of 4-4.5 compared to the initial state).

CONCLUSIONS

We showed that the sequential double implantation of N^+ and Ni^+ ions into nitinol (TiNi) leads to the formation of a complex depth profile of the nitrogen concentration, which is caused by the rejection of nitrogen ions from the region of the maximum Ni^+ ion losses (i.e., the region of the maximum Ni^+ ion concentration) to the region of residual tensile stresses. As a result of the implantation of N^+, Ni^+, W^+, and Mo^+ ions, the hardness after implantation increases by 30% compared to the initial state and the hardness after subsequent thermal annealing at 550°C for 2 h increases by a factor of 2.2. The SME changes because of the formation of nitrogen and carbon (carbonitride) layers as a result of N^+ implantation and because of a change in the concentrations of Ti and Ni atoms due to the sputtering of Ni atoms, which is accompanied by a change in the martensite transformation temperature.

The martensitic form of equiatomic NiTi was implanted with N ions with the fluence 10^{18} cm^{-2} and 50 keV energy. To characterize the transformation sequence and transformation temperatures, the DSC measurements were carried out on an unimplanted as well as an implanted material. Both the unimplanted and ion-implanted NiTi alloys transform in two steps ($B_2 \rightarrow R \rightarrow B_{19}$) in the cooling direction and one-step transition ($B_{19} \rightarrow B_2$) in the heating process. To verify identifications of martensitic transformations in the NiTi alloys during heating and cooling, the X-ray structural investigations were performed.

The TEM structural characterization reveals the existence of the well-defined double-layer structure with different microstructure as well as different phase and chemical composition in the near-surface region of the ion-implanted NiTi alloy. From the surface to a depth of 80 nm, the sample has an amorphized structure in the form of two sublayers: the first is a Ti- and N-rich nano-crystalline and/or amorphous-like and the second - Ni-rich crystalline. In the depth of 80-160 nm the material has a defected Ti-rich crystalline microstructure and deeper - an unaffected grain structure of the parent material.

REFERENCES
[1]Treatise on .Materials Science and Technology, Ed. by W. F. Wang, Vol. 18: Ion Implantation. Ed. by J. K. Hirvonen. *Academic, New York (1980); Metallurgiya, Moscow (1985)*.
[2]A.D. Pogrebnjak, A.P. Kobzev, B.P. Gritsenko et al., Effect of Fe and Zr ion implantation and high-current electron irradiation treatment on chemical and mechanical properties of Ti-V-Al alloy, *J. of Applied Physics*, **87**, 5, 2142–48 (2000).
[3]E.A.Levashov, D. V. Shtansky, Multifunctional nanostructured films, *Russian Chemical Reviews*, **76** (5), 463-470 (2007)
[4]V. M. Anishchik, V. V. Uglov, Ion Implantation into Tool Steel, *BGU. Minsk.* (2000) [in Russian].
[5]A.M.Perez-Martin, A.M. Vredenberg, L. de Wit et al., Carbide and nitride/carbide layers in iron synthesized by ion implantation, *Mater. Sci. and Eng.*, **B19**, 281–84 (1993)
[6]D.Chrobak, H.Morawiec, Thermodynamic analysis of the martensitic transformation of plastically deformed NiTi alloy, *Scr. Mater.*, **44**, 725–30 (2001).
[7]S.Shabalovskaya, J.Andregg, J.Van Humbeeck, Critscal overview of Nitinol their modifications for medical appltcations, *Acta Biomater*, **4**, 447–67 (2008).
[8]Su-Young Cha, Jeong Se-Young Jeonc, Joung Hum Park et al., Thermodynamic and structural characterization of high – and low – temperature nitional, *J Korean Phys. Soc.*, **49**, S580–83 (2006).
[9]K.K. Kadyrzhanov, F.F., Komarov, A.D. Pogrebnjak etc., Ion-Beam and Ion-Plasma Treatment of Materials. *Moscow: Moscow State Univ.*, 640p. (2005).
[10]A.D. Pogrebnjak, S.N. Bratushka, V.V. Uglov et al. Structure and properties of Ti alloys after double implantation, *Vacuum*, **83**, 6, S241–44 (2009).

[11]D.M. Shirokov, V. Bohac, New computer iterative fitting program DVBS for backscattering analysis, *Nucl. Instr. and Methods in Psys. Resear.*, **84(B),** 497–506 (1994).

[12]T.Czeppe, N. Levintant-Zayonts, Z. Swiatek, et al., Inhomogeneous structure of near-surface layers in the ion-implanted NiTi alloy, *Vacuum*, **83**, S214-19 (2009).

[13]A. D. Pogrebnyak, S. N. Bratushka, N. Levintanr, et al., Effect of High Doses of N^+, $N^+ + Ni^+$, and $Mo^+ + W^+$ Ions on the Physicomechanical Properties of TiNi, *Technical Physics, 54, No. 5, 667-73* (2009).

[14]A.D. Pogrebnjak, A.P. Kobzev, B.P.Gritsenko et.al., Effect of Fe and Zr ion implantation and high-current electron irradiation treatment on chemical and mechanical properties of Ti-V-Al alloys, *Jour. of Appl.Phys*, **87**, 5, 2142-47 (2000).

[15]A.D. Pogrebnjak, E.A. Bazyl', Modification of wear and fatigue characteristics of Ti-V-Al alloy by Cu and Ni ion implantation and high-current electron beam treatment, *Vacuum*, **64**, 1, 1-7 (2001).

[16]A.D.Pogrebnjak, O.G.Bakharev, N.A.Pogrebnjak et.al., Certain features of high-dose and intensive implantation of Al ions in iron, *Phys.Lett.*, **A265**, 3, 225-32 (2000).

DESIGN AND CONSTRUCTION OF COMPLEX NANOSTRUCTED AL$_2$O$_3$ COATING FOR PROTECTIVE APPLICATIONS

P. Manivasakan and V. Rajendran*
Centre for Nano Science and Technology, K. S. Rangasamy college of Technology,
Tiruchengode-637 215, Tamil Nadu, India
*Corresponding author: veerajendran@gmail.com

P. R. Rauta, B. B. Sahu and B. K. Panda
Dalmia Institute of Scientific and Industrial Research, Rajgangpur-770 017, Orissa, India

ABSTRACT

In the present investigation, alumina nanopowder was obtained by spray pyrolysis using an inexpensive precursor of aluminum nitrate synthesized from raw bauxite. The powders produced were comprehensively characterised employing X-ray diffraction, particle size distribution, Fourier transform infrared spectroscopy, Brunauer- Emmett-Teller surface area and pore size analysis, X-ray fluorescence spectrometry, and transmission electron microscopy studies. The nanopowder of -Al$_2$O$_3$ with an average crystallite size of 5 nm and an average particle size of 28 nm with a specific surface area of 336 m^2 g^{-1} was produced. The obtained result reveals that the obtained nano powders were having spherical morphology with free flowing structure. Design and construction of complex nano structured Al$_2$O$_3$ coating on stain steel specimen was performed by a dip-coating technique. The thickness of nano- Al$_2$O$_3$ engineered silica film coating on steel specimen was controlled and varied by using layer-by-layer coating method. The effect of nano-coating on anti-corrosive properties of stain steel in acid media was performed using conventional weight loss method.

1. INTRODUCTION

The protective coatings are used for wide variety of applications[1] like cars, houses and bridges to cover and protect underlying material. An individual coating layer may contain many components such as organic or inorganic binder, a liquid solvent to dissolve or suspend components and a variety of additives such as surfactant, plasticizer, biocides, cross linking agent, surface particles to control reflectivity and transport etc. Multilayer coatings form complexity of the thin film on final product. Typical coating products are manufactured and sold with several coating layers. A thin surface layer of ceramics deposited on metal imparts favorable ceramic characteristics such as corrosion resistance and wear resistance while retaining the durability and structural benefits of the metal[1].

The production of Al$_2$O$_3$ nanoparticles has been intensively pursued due to its many technological applications such as coatings, heat-resistant materials, abrasive grains, coated/super abrasives, cutting materials, advanced ceramics and an additive of paints and pigments[2]. This is mainly because of its important physico-chemical properties such as hardness, high chemical and thermal resistance, dimensional stability up to 1500 °C and an excellent mechanical strength and wear resistance[3]. By comparing micron-sized and nano-sized alumina particles, it has been found that nano-alumina has many advantages due to its high surface area[4] at nano-scale range. In view of the unique properties, such as high surface area, porosity and chemical activity, the nano-alumina particles are used for high-temperature applications, adsorbents, coatings and soft abrasives[5]. The use of nano-sized alumina particles has significantly improved the quality and reproducibility of fine coatings[6]. The development of inexpensive methods for large-scale production of high surface area Al_2O_3 nanoparticles remains an area of extensive interest. Various synthesis methods are being explored and developed for the production of ultra-fine alumina particles through solution-based techniques such as sol-gel[7], hydrothermal[8], microwave[9] and micro-emulsions[10]. However, most of these methods are proposed from chemically available precursors that are highly expensive for bulk production. It is important to note that the production of Al_2O_3 nanoparticles in large quantity is essentially required for their extensive industrial applications such as surface productive coatings, ceramics and refractories. The production of alumina nanopowder from raw bauxite by the sol-gel method has been reported by us previously[11].

In this article, we report the production of non-aggregated nano-Al_2O_3 particles employing spray pyrolysis using aluminium nitrate obtained from natural bauxite. A review of literature proves that this is an innovative first-time effort for the production of γ-alumina nanoparticles from raw bauxite. Producing nano-Al_2O_3 particles from inorganic salt solutions is an inexpensive method for large-scale production. The successful synthesis of uniform-sized cubic γ-alumina particles with an average crystallite size of 5 nm and an average particle size of 28 nm with a specific surface area (SSA) of 336 $m^2\ g^{-1}$ was produced. Nano-scale particles with free of coarse agglomerates have been employed for several fabrication techniques which includes colloidal spray deposition, dip coating, slip-casting, tape casting and isostatic pressing[12]. Design and construction of complex nano- structured Al_2O_3 protective coating on the surface of stainless steel (SS 304) has been performed by a dip-coating technique using nano- Al_2O_3 engineered silica sol (AES). The effects of coating and coating thickness have been monitored using anti-corrosion studies of stainless steel in acid media. The present paper was aimed at investigating the effect of AES coating on the corrosion resistance of SS 304 in acid media.

2. EXPERIMENTAL

2.1 Production of Al₂O₃ nanoparticles

a) Preparation of aluminium nitrate precursor

The commercial raw bauxite, which contains 50–55% of Al_2O_3 obtained from the Chhatrapur region of the coastal part of Orissa, India was used as a starting material. The dry mixture which consists of 66.67 wt% of raw bauxite and 33.33 wt% of sodium hydroxide (99.9%; Merck GR) was fused at 600 °C for 3 h in a muffle furnace. The resultant product was leached with double-distilled water. The obtained solution mixture was adjusted to a pH value of 13±0.5 using a 5 N sodium hydroxide solution followed by vigorous magnetic stirring at 353 K for 1 h. The stirred solution was filtered and the filtrate was precipitated with 6 M HNO_3(69%, Merck GR) solution while being constantly stirred and an amorphous hydrated aluminium oxide was precipitated at pH 7±0.5. The obtained precipitate was washed with double distilled water. Further, the precipitate was dissolved in concentrated nitric acid to get aluminium nitrate.

b) Experimental set-up

The block diagram of automated spray pyrolysis experimental set-up is shown in Fig. 1. The spray pyrolyser experimental set-up primarily consist of i) an atomiser which converts the starting solution into droplets, ii) automated anti-blocking unit, iii) a tubular electric furnace with hot air blower, iv) two-fluid nozzle with compressed air inlet and sample feeding port, v) feed pump which facilitate the flow rate of precursor, vi) reaction chamber, vii) cyclonic sample collectors and viii) purification system. The total automated experimental set-up is controlled through a single control panel.

c) Spray pyrolysis

The true homogeneous solution of aluminium nitrate precursor was used as the starting phase in spray-pyrolysis to obtain nano Al_2O_3 particles. In spray-pyrolysis, reaction often takes place in solution in droplets followed by solvent evaporation. The method is based on atomising the precursor and injecting the spray into a tubular reaction chamber. The atomised droplets of the precursor are converted into nano sized entities during their flow through the tubular reaction chamber. The hot air is introduced into the reaction chamber followed by the precursors are sprayed into chamber with use of two-fluid nozzle pressurised with compressor air. The feed pump is used to control the flow rate of precursors. The formation of atomiser is controlled by controlling the pressure of compressed air. The sprayed and atomised nano sized entities of $Al(NO_3)_3$ were decomposed at 523 K to obtain nano Al_2O_3 particles. After the completion of one full cycle, the produced nano Al_2O_3 particles were collected from the cyclones.

2.2 Preparation of nano- Al_2O_3 engineered silica sol

The silica sol was prepared from tetraethyl orthosilicate (TEOS; Merck; 99%), Conc.HNO_3 and ethanol. 2.5 ml of TEOS (0.1M) and 90 ml of ethanol were mixed thoroughly for 10 minutes and add 5 ml of polyethylene glycol to slow down the solvent evaporation. 2.5 ml of 69 % HNO_3 was then added dropwise with continuous stirring for 20 minutes and stirring was continued for 1 h. It was followed by 0.5 g of as-produced nano Al_2O_3 particles were dispersed in silica sol under sonication for 30 minutes.

2.3 Dip coating

Dip coating was performed in the stable solution of nano- Al_2O_3 engineered silica sol. The stainless substrates were coated by dipping and withdrawing from the solution at a constant speed of about 1 mm s^{-1}. The coated substrates were allowed to dry for 1 h at ambient temperature. The coating was then heat treated by firing at 400±25°C for 0.5 h followed by a second stage firing at 800±25°C for 0.5 h using the heating rate of 5°C/ min and then allowed for furnace cooling. In order to deposit relatively thick coatings it is necessary to build up multiple layers with each layer fired separately to avoid cracking and delimitation. The stainless steel substrate coated with 1, 3 and 6 layers of nano-Al_2O_3 engineered silica (here after named as 1AES, 3AES and 6AES) were prepared in the present study. Gravimetric analysis was also used to evaluate the thermal degradation of the coatings.

2.4 Corrosion studies

The coated and uncoated stainless steel specimens were immersed in 1M HCl solution for 1 day to determine the effect of nano-coating against the corrosion of stainless steel 304 in acid media.

The conventional weight loss method was employed to obtain the corrosion loss of stainless steel. The percentage corrosion inhibition efficiency was calculated using the following equation,

$$\% \ IE = \frac{W_o - W}{W_o} \times 100$$

where W_o and W are the weight loss of uncoated and coated SS plate respectively. The corrosion rate was calculated using the following formula,

$$Corrosion \ rate = \frac{Weight \ loss \ in \ mg}{Surface \ area \ in \ dm^2 \ x \ immersion \ period \ in \ days}$$

2.4 Characterisation

X-ray powder diffraction (XRD) patterns were obtained by X'pert pro, PANalytical, GG Eindhoven, Netherland, using CuKα as a radiation source at 40 kV and 30 mA. Infrared spectra (IR) were recorded using FTIR spectrophotometer (Spectrum 100, Perkin Elmer, Waltham, MA, USA) employing KBr pellet method (95 Wt.% KBr). The chemical purity of the as-synthesised Al$_2$O$_3$ particles was determined through wet chemical analysis by the EDTA titration method according to the Indian Standard method [IS 1760 (Part 3):1992]. Particle size distribution was measured using Nanophox particle size analyser, Sympatec, Clausthal-Zellerfeld, Germany. The specific surface area of the powdered samples was determined employing the BET technique using an Autosorp AS-1MP system, Quantachrome, Boynton Beach, FL, USA. Qualitative and quantitative elemental analysis of alumina particles were performed using XRF spectrometry (EDX-720, Shimadzu, Kyoto, Japan). Transmission electron microscopy (TEM) pictures were obtained using a Philips electron microscope, CM 200, Hillsboro, OR, USA. Elemental analysis of coated and un-coated stainless steel specimen was performed using arc/spark optical emission spectrometry (ARL 4460, Thermo Scientific, Waltham, MA, USA). Surface topographic images of coated and uncoated stainless steel specimens were performed using scanning electron microscope (JSM - 6390LV, JEOL, Tokyo, Japan). The coating thickness was measured using gravimetric method[13].

3. RESULTS AND DISCUSSION

Representative XRD pattern of Al$_2$O$_3$ particles is shown in Fig. 2(a). The diffraction pattern of Al$_2$O$_3$ particles is assigned to the cubic phase, indicating that the Al$_2$O$_3$ particles has crystalline γ-Al$_2$O$_3$ phase with cubic symmetry. The obtained XRD pattern of Al$_2$O$_3$ particles is agreed well with the standard powder diffraction data (JCPDS File No.: 79-1558). From the XRD data, it is found that Al$_2$O$_3$ particles has cubic crystalline phase with an average crystallite size of 3 nm. Fig. 2(b) shows the FTIR spectra of Al$_2$O$_3$ particles after calcination at 500 °C. The band observed at 1633 cm^1 is assigned to the O–H bending vibration of a weakly bounded water molecule[14]. The absorption peaks obtained at 1510 and 1410 cm^1 are due to the presence of the carbon–carbon (C–C) deformation. The band observed at 1100 cm^1 is assigned to the Al–OH bending vibration of Al–OH–Al groups[14]. The absorption peaks observed between 600 and 800 cm^1 are attributed due to the stretching and librational modes of the AlO_4, AlO_6 and OH groups[14]. The band at 502 cm^1 is assigned to the stretching and bending vibrations of the Al–O bond[14]. The above results summarise the finding that Al_2O_3 particles consists of adsorbed water molecules and residual carbon.

Chemical analysis shows that as-produced sample contains 98.9% of Al_2O_3. It can be seen from XRF (Table I) results that the sample contains 99.2% of Al_2O_3. The results obtained from quantitative elemental analysis (XRF) are in close agreement with the results obtained from the chemical analysis. From the above studies, it was concluded that as-produced Al_2O_3 particles shows 98.9% chemical purity. Fig. 2(c) presents the PSD of as-received Al_2O_3 particles after calcination at 500 °C. It can be seen from Fig. 2(c) that the synthesised powder consists of particle size in the range of 12-51 nm and the maximum distribution of particle is at 25 nm. The surface area of the resulting nano-Al_2O_3 particles calculated using BET method is 336 m^2 g^{-1}. The nano-Al_2O_3 particles maintained larger surface area above 200 m^2 g^{-1} even after calcination at 1023 K. It can be revealed from BET analysis that the spray pyrolysis yields high surface particles with free flowing structure. Further, it is evident from the above studies that the particles are freely functionalized and easily fabricate to different nanostructures depending on the requirements. The TEM image of Al_2O_3 particles after calcination at 500 °C is shown in Fig. 2(d). From the figure, the morphologies and particle size of Al_2O_3 particles can be confirmed. It can be seen from the TEM images that the Al_2O_3 particles exhibit with monodispersed spherical particles with an average particle size of 25 nm. The observed results confirm that the particles obtained from raw bauxite have uniform particle size distribution with monodisbursed spherical morphology.

Table II shows that the elemental composition of uncoated and nano- Al_2O_3 engineered silica coated stainless steel specimens. It was confirmed from spark optical emission studies (Table II) that the uncoated stainless steel specimen is SS 304. It was observed from Table II that nano- Al_2O_3 engineered silica coated SS 304 shows an increased amount of Al and Si composition and remaining elements are almost unaltered. It is used to frame the discussion that the increase in composition of Al and Si confirms the presence of complex nanostructed Al_2O_3 protective coating on SS 304. Nano-Al_2O_3 engineered silica film coated on the surface of the stain less steel 304 isolates the underlying metal from the corroding environment. Table III shows the percentage inhibition efficiency and corrosion rate of coated and uncoated stainless steel 304 in hydrochloric acid media. It was found that an increase in coating thickness leads to a decrease in H$^+$ ions diffusion over the SS 304 substrate which inturn increases the inhibition efficiency and decreases its corrosion rate. It can be seen from Table III that nano- Al_2O_3 engineered silica sol coating shows better acid corrosion resistance than silica sol coated SS 304. This may be due to the pin holes present in the silica film network on SS surface which was arrested by dispersion of nano- Al_2O_3 particles in silica matrix. In addition, incorporation of nano- Al_2O_3 particles in silica matrix induces better densification of silica protective film over the SS 304 substrate.

The above studies reveal that nano- Al_2O_3 engineered silica film is acting as a barrier to block the H$^+$ ion diffusion over the stainless steel surface. Therefore, the coated barrier reduces the H_2 evolution and inturn control the release of Fe^{2+} ions when it's in contact with acid solution. Further, it is noted that an increase in coating thickness increases the protection rate against H$^+$ ion diffusion over the stainless steel surface. The multilayer coating shows an effective corrosion barrier against HCl corrosion when compared to single layer coating. The SEM micrograph of polished SS 304 (Fig.4a) reveals that the surface is uniform and the parallel features seen on the surface may be associated with polishing scratches while in 1 M HCl, SEM micrograph (Fig.4c) reveals that the surface was covered with high density of pits. It can be observed from Figures 4a and 4c that the acid solution effectively damages the SS 304 specimen due to the effective diffusion of H$^+$ ions over SS 304 substrate. Figures 4b and 4d show the SEM micrograph of nano- Al$_2$O$_3$ engineered silica coated SS 304 before and after immersion in 1 M HCl. It is observed from Figure 4b that the dip coating process leads to almost uniform coating on SS 304 surface. It can be seen from Figure 4d that the complex nanostructed Al$_2$O$_3$ coating efficiently protects H$^+$ ion diffusion over the stainless steel surface in HCl media. Further, SEM observation is in close agreement with corrosion studies that nano- Al$_2$O$_3$ engineered silica coating shows better corrosion barrier against acid (1M HCl) corrosion of SS 304.

4. CONCLUSION

An inexpensive and eco-friendly method for production of Al$_2$O$_3$ nanoparticles has been developed using raw bauxite directly. The size characterisation was performed with XRD, PSA, TEM and BET studies. Highly spherical cubic nano-alumina (-Al$_2$O$_3$) particles with an average particle (d$_{50}$) size of 25 nm and a specific surface area of 336 m^2 g^{-1} were produced from raw bauxite through spray pyrolyser. The possibilities for large scale production of nano- Al$_2$O$_3$ particles from bauxite were investigated. The nano- Al$_2$O$_3$ engineered silica sol was coated on SS 304 surface for the protection of acid corrosion. It was revealed from corrosion and microscopic studies that the nano- Al$_2$O$_3$ engineered silica coating shows better corrosion barrier than the silica sol coated and uncoated SS 304 against acid (1M HCl) corrosion at room temperature. It was found that an increase in coating thickness leads to a decrease in H$^+$ ions diffusion over the SS 304 substrate which inturn increases the inhibition efficiency and hence decreases its corrosion rate.

ACKNOWLEDGEMENT

The authors are very much thankful to Department of Science and Technology, New Delhi for the financial support to carry out this research project under Grant No. SR/S5/NM-40/2005 dt. 26.06.07.

REFERENCES

1. X. Miao and B. Ben-nissan, "Microstructure and properties of zirconia-alumina nanolaminate sol-gel coatings", *Journal of Materials Science,* **35** 497– 502 (2000).

2. T. B. Du, S. M. Jang and B. W. Chen, " Manufacture of mesoporous alumina of boehmite type via subcritical drying and application to purify liquid crystal", *Chem. Engi. Sci.,* **62** 4864-4868 (2007).

3. J. Tikkanen, K. A. Gross, C. C. Berndt, V. Pitkanan, J. Keskinen, S. Raghu, M. Rajala and J. Karthikeyan, "Characteristics of the liquid flame spray process", *Surf. Coat. Technol.,* **90** 210-216 (1997).

4. A. I. Y. Tok, F. Y. C. Boey and X. L. Zhao, "Novel synthesis of Al_2O_3 nano-particles by flame spray pyrolysis", *J. Mater. Proc.Technol.,* **178** 270-273 (2006).

5. S. Kim, Y. Lee, K. Jun, J. Park and H. S. Potdar, "Synthesis of thermo-stable high surface area alumina powder from sol–gel derived boehmite", *Mater. Chem. Phys.,* **104** 56-61 (2007).

6. Y. Q. Wu, Y. Zhang, X. Huang and J. Guo, "Preparation of plate like nano alpha alumina particles", *Ceram. Int.,* **27** 265-268 (2001).

7. S. Kim, Y. Lee, K. Jun, J. Park and H.S. Potdar, "Synthesis of thermo-stable high surface area alumina powder from sol–gel derived boehmite", *Mater. Chem. Phys.,* **104** 56-61 (2007).

8. C. Kaya, J. Y. He, X. Gu and E. G. Butler, "Nanostructured ceramic powders by hydrothermal synthesis and their applications", *Microporous and Mesoporous Mater.,* **54** 37-49 (2002).

9. S. G. Deng and Y. S. Lin, "Microwave synthesis of mesoporous and microporous alumina powders", *J. Mater. Sci. Lett.,* **16** 1291-1294 (1997).

10. Y. Pang and X. Bao, "Aluminium oxide nanoparticles prepared by water-in-oil microemulsions", *J. Mater. Chem.,* **12,** 3699-3704 (2002).

11. P. Manivasakan, V. Rajendran, P. R. Rauta, B. B. Sahu and B. K. Panda, "Direct synthesis of nano alumina from natural bauxite", *Advanced Materials Research* **67** 143-148 (2009).

12. G. Xu, Y. W. Zhang, C. S. Liao and C. H.Yan, "Grain size-dependent electrical conductivity in scandia-stabilized zirconia prepared by a mild urea-based hydrothermal method", *Solid State Ionics,* **166 [3-4]** 391-396 (2004)

13. A.U. Ubalea, V. P. Deshpandea, Y. P. Shindea and D. P. Gulwadeb, "Eectrical, optical and structural properties of nanostructured Sb_2S_3 thin films deposited by CBD technique", *Chalcogenide Letters,* **7[1],** 101 – 109 (2010).

14. G. L. Teoh, K. Y. Liew and W. A. K. Mahmood, "Synthesis and characterization of sol-gel alumina nanofibers", *J. Sol-Gel Sci. Technol.,* **44** 177-186 (2007).

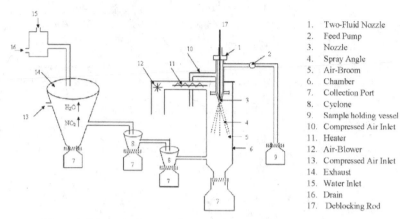

1.	Two-Fluid Nozzle
2.	Feed Pump
3.	Nozzle
4.	Spray Angle
5.	Air-Broom
6.	Chamber
7.	Collection Port
8.	Cyclone
9.	Sample holding vessel
10.	Compressed Air Inlet
11.	Heater
12.	Air-Blower
13.	Compressed Air Inlet
14.	Exhaust
15.	Water Inlet
16.	Drain
17.	Deblocking Rod

Figure 1. Schematic diagram of automated spray pyrolyser experimental set-up

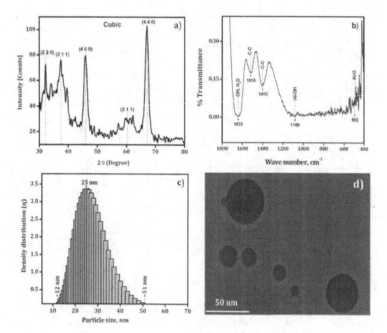

Figure 2. (a) XRD pattern of Al$_2$O$_3$ particles, (b) FTIR spectra of Al$_2$O$_3$ particles, (c) Particle size distribution of Al$_2$O$_3$ particles and (d) TEM image of Al$_2$O$_3$ particles

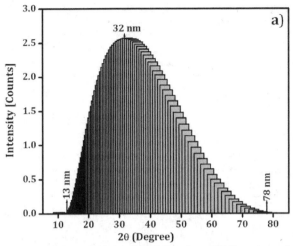

Figure 3. Particle size distribution of nano-Al₂O₃ engineered silica sol

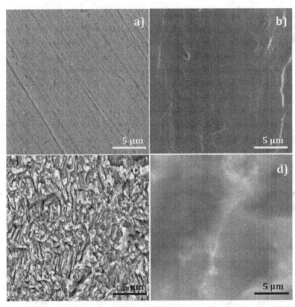

Figure 4. SEM micrograph of (a) polished SS 304 surface, (b) nano-Al₂O₃ engineered silica coated SS 304, (c) uncoated SS304 in 1M HCl and (d) Al₂O₃ engineered silica coated SS 304 in1M HCl.

Table I Chemical Composition of Al_2O_3 particles

XRF chemical composition	
Components	Wt % (± 0.01)
Al_2O_3	99. 20
Na_2O	0.25
SiO_2	0.54
CaO	0.01

Table II Elemental analysis of uncoated and coated stainless steel

Uncoated Stainless Steel Specimen		Coated Stainless Steel Specimen	
Elements	Composition (%)	Elements	Composition (%)
C	0.05	C	0.05
Si	0.47	Si	1.67
Mn	1.14	Mn	1.14
P	0.037	P	0.035
S	0.011	S	0.01
Ni	8.53	Ni	8.53
Cr	18.92	Cr	18.92
Mo	0.22	Mo	0.22
V	0.12	V	0.12
Cu	0.15	Cu	0.15
W	0.013	W	0.013
Ti	0.021	Ti	0.021
Sn	0.02	Sn	0.02
Co	0.13	Co	0.13
Al	0.004	Al	0.52
Pb	0.02	Pb	0.02
B	0.001	B	0.001
Sb	0.092	Sb	0.092
Nb	0.034	Nb	0.034
Zr	0.022	Zr	0.022
Bi	0.051	Bi	0.051
Ca	0.001	Ca	0.001
Mg	0.005	Mg	0.005
Zn	0.013	Zn	0.013
Ce	0.002	Ce	0.002
La	0.008	La	0.008
Fe	69.78	Fe	67.56

Table III Coating thickness, % Inhibition efficiency

SS 304 Specimen Coatings	Number of Layers	Thickness μm (± 0.1)	% Inhibition efficiency (± 5)	Corrosion rate mg dm^{-2}day^{-1}(mdd)
without coating	0	-	-	256
silica sol	1	0.2	40	154
	3	0.5	63	95
	6	0.8	70	77
nano-Al$_2$O$_3$	1	0.3	60	102
engineered silica sol	3	0.7	88	31
	6	0.9	92	20

MICROWAVE ASSISTED SYNTHESIS AND CHARACTERIZATION OF SILVER/GOLD NANOPARTICLES LOADED CARBON NANOTUBES

Charity Maepa[a,b] *, Sreejarani K. Pillai[a], Suprakas Sinha Ray[a], Leskey Cele[b]

[a]DST/CSIR Nanotechnology Innovation Centre, National centre for Nano-Structured Materials, Council for Scientific and Industrial Research, Pretoria 0001, South Africa
[b]Department of Chemistry, Tshwane University of Technology, Pretoria 0001, South Africa
*Corresponding author. E-mail: CMaepa@csir.co.za

ABSTRACT

Microwave assisted synthesis of nanostructured materials has recently shown remarkable advantages over the conventional synthesis routes such as rapid volumetric heating, high reaction rate, reduced particle size, homogeneous and narrow size distribution of particles. In this work, we report a comparison between conventional and microwave assisted methods for loading of silver-gold (Ag-Au) nanoparticles on the surface oxidized carbon nanotubes (CNTs). Both wet impregnation and deposition precipitation methods were used as the conventional methods for the preparation of the Ag-Au/CNT heterostructures. The synthesized heterostructures were Characterized using high-resolution scanning and transmission electron microscopy, Raman spectroscopy and X-ray diffraction. Results showed that microwave assisted- synthesis method produced much better distribution and property enhancement on Ag-Au/CNT composites when compared to the conventional methods.

INTRODUCTION

Many methods have been discovered for the synthesis of the intriguing carbon nanotubes (CNTs) in their different type such as single, double and multi walled carbon nanotubes [1-7]. These materials have been studied extensively since their discovery almost two decades ago and have shown remarkable properties that includes its outstanding mechanical properties, electronic and thermal conductivities which thus opened an opportunity for various potential applications in catalysis [1-8], sensors [9-12], composites [13], batteries, field emission displays, etc [1-7, 9-13]. These synthetic methods, all with the goal of achieving high quality and quantity of the CNTs, includes the Arc discharge, laser ablation, pyrolysis and chemical vapour deposition (plasma enhanced , thermal and catalytic) [1-7, 14, 15]. The chemical vapour deposition (CVD) was used to synthesize the multi-walled CNTs (MWCNTs) in this study, since the CVD synthesis is outstanding not only because of large scale production of the CNTs but also its good quality, versatility and low cost [14, 15].

The decoration of the CNTs with metal nanoparticles such as Pt, Pd, Ag and Au has also shown an increasing interest worldwide since the composite materials [8, 16-26], in contrast to their counterparts exhibits unique properties. The bimetallic combination of these nanoparticles especially are capable of forming an alloy when reduced simultaneously [21]. These materials exhibit abnormal non-linear optical limiting behaviour which enables them to be applied for coatings, electronic devices, gas sensing, environmental and industrial catalysis. The interest is thereof studying the combination of transition metals with the CNTs in order to enhance their properties and applications [10, 12, 27]. The common synthetic methods used to achieve the Ag-Au loaded CNT heterostructures involves reduction

103

of the corresponding metals metal salt by using ethanol as a reducing agent under refluxing conditions. The laser ablation method is another synthetic route as reported by Paszti et al. and the challenge of particle agglomeration and lack of particle dispersion remains a problem for both the methods [10, 12, 27]. The other conventional methods involve the wet impregnation, deposition precipitation and may other reported methods. However, to our knowledge the use of microwave method for the synthesis of bimetallic Ag and Au nanoparticles decoration on CNTs has not been reported.

In this work, we present a comparative study on the characteristics of bimetallic nanoparticles of Ag and Au on the CNTs synthesized using different methods. Apart from the conventional methods of synthesis like wet impregnation and deposition precipitation, we used microwave assisted wet impregnation as a new technique to synthesize the composite materials. What is special about this technique is its advantage on rapid volumetric heating, enhanced reaction kinetics, reduced particle size, homogeneous and narrow size distribution of particles as compared to the conventional methods [10, 12, 27]. The heterostructures thus prepared were studied using high resolution transmission electron microscopy (HR-TEM), scanning electron microscopy (SEM), energy dispersive X-ray (EDS), Raman spectroscopy and X-ray diffraction (XRD) analysis. The results showed that microwave assisted-synthesis method produced much better distribution and property enhancement on the Ag/Au/CNT structures when compared to other conventional methods.

EXPERIMENTAL

Catalyst preparation for CNT synthesis

$Fe(NO_3)_3 \cdot 9H_2O$ (1.8 g) and $Co(NO_3)_3$. $6H_2O$ (1.23g) were dissolved in 60 ml of de-ionized water and then impregnated on Calcium Carbonate (10g) which is used as a support. The mixture was stirred for an hour, filtered and dried in the oven for 16 hours at 120°C. The dried catalyst was ground and sieved, then calcined in air at 400°C with a heating rate of 5°C /min for 16 h.

Synthesis of MWCNTs

CNTs were synthesized by the decomposition of acetylene (C_2H_2) (Afrox) in a tubular quartz reactor (150 cm × 4.5 cm i.d.) that was placed horizontally in a furnace. The catalyst (0.7 g) was spread to form a thin layer in a quartz boat (120 mm × 15 mm) and the boat was then placed in the centre of the reactor quartz tube. The furnace was then heated at −13 °C/min under flowing N_2 (40 ml/min). Once the temperature had reached 675 C, the N_2 flow rate was set to 240 ml/min and C_2H_2 was introduced at a constant flow rate of 90 ml/min. After 30 min of reaction time, the C_2H_2 flow was stopped and the furnace was left to cool down to room temperature under a continuous flow of N_2 (40 ml/min). The boat was then removed from the reactor and the carbon deposit that formed along with the catalyst was weighed. The typical yield of CNTs obtained from 0.7 g of catalyst used was 2.0 g.

Purification and oxidation of CNTs

To isolate the CNTs from the CNT/catalyst mixture, about 7g of the mixture was washed with 30% HNO_3 by stirring it for about 1 h at room temperature and dried at 110°C for overnight in an oven. The purified CNTs were then oxidized by reflux in 30% HNO_3 for 2h at 120°C, filtered, dried and used as a support for metal loading.

Ag-Au-loading on CNTs

Microwave method: 200 mg of oxidized CNTs were added to a mixture of solutions of $AgNO_3$ (50 ml) and $AuCl_4H$.aq (50 ml) (calculated for 2.5 wt% of Ag and 2.5 wt% of Au) and the loading of the metal on to CNTs was done in microwave reactor (Perkin Elmer microwave reaction system-Multiwave 3000) at a power setting of 500 W for 10 min at 200°C. The precipitate was then removed by centrifugation and washed several times with distilled water (until the supernatant liquid was free from NO_3^- and Cl^- ions) and dried at 110°C for overnight.

Wet Impregnation: 200 mg of oxidized CNTs were added to the solution mixture of $AgNO_3$ and $AuCl_4H$.aq (calculate for 2.5 wt% of Ag and 2.5 wt% of Au) and sonicated for 3-5 min with high power ultrasonic finger and magnetically stirred for 2 h at room temperature. The precipitate was then removed by centrifugation and washed several times with distilled water (until the supernatant liquid is free from NO_3^- and Cl^- ions) and dried at 110°C overnight.

Deposition precipitation: A desired amount of oxidized CNTs and Ag and Au salts are mixed with de-ionized water, and a diluted solution of NH_4OH was then added slowly under vigorous stirring until the solution reached pH 7. The loaded CNTs were filtered and washed thoroughly with distilled water, dried at 110°C overnight.

The sample notations used for different samples were CNT (pure), Ag-Au/CNT-WI (wet impregnation), Ag-Au/CNT-DP (deposition precipitation), Ag-Au/CNT-MW (microwave synthesis).

CHARACTERIZATION

The surface morphology of the samples were analysed by JEOL7500F field-emission SEM (FE-SEM) at an accelerating voltage of 2 kV. The powder sample was mounted on the copper stub using a carbon tape. The samples were sputter coated by carbon to avoid charging. The elemental analysis was done with EDS system attached with the SEM instrument. The crystalline phases of the samples were determined by X-ray diffraction (PAnalytical XPERT-PRO diffractometer) measurement, using Ni filtered CuKα radiation (λ = 1.5406 Å), with variable slit at 35 kV, 50 mA. The structure of Ag-Au loaded CNTs composites were analyzed by TEM (JEOL, JEM-2100). For TEM, the CNTs were dispersed in the ethanol by ultrasonication to form a suspension. One or two droplets were dropped onto holey carbon supported copper grid and allowed to dry. Raman spectra of samples were recorded by a Raman spectroscopy (Jobin-Yvon T64000 spectroscope), equipped with an Olympus BX-40 microscope attachment. The excitation wavelength was of 514 nm with energy setting 1.2 mW from a Coherent Innova Model 308 Argon ion laser.

RESULTS AND DISCUSSION

SEM

SEM images of pure and Ag-Au modified CNTs are given in figure 1 (a) (d). The purified CNTs are quite uniform in size, i.e. in the range 25–35 nm in diameter (refer Figure 1a). Dense clusters of randomly oriented (spaghetti-like) CNTs with high aspect ratios were observed. The SEM image of the unmodified CNTs reveals that the oxidation process did not cause any fragmentation and agglomeration of the CNTs. The image shows good quality tubes with no amorphous carbon or catalyst support.

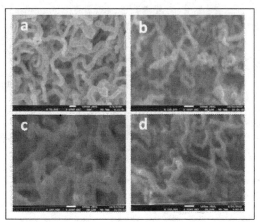

Figure 1. SEM micrographs of (a) CNT (pure), (b) Ag-Au/CNT-WI (wet impregnation), (c) Ag-Au/CNT-DP (deposition precipitation), (d) Ag-Au/CNT-MW (microwave synthesis).

SEM images of the CNTs modified by Au-Ag metals by different methods are shown in Figure 1 (b) (d). The images of the functionalized samples show nanotubes with open ended tips. The tips are opened due to surface oxidation induced by HNO_3. It is worthy to note that the CNTs exhibit no obvious length decrease or damage after surface functionalization. On the other hand, the roughening of the tubes is observed to be more for the modified samples which may be due to the deposition of metal nanoparticles on the surface.

EDS

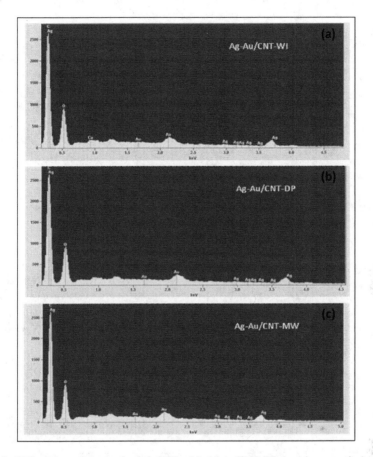

Figure 2. EDS spectra of Ag-Au loaded CNTs: (a) Ag-Au/CNT-WI (wet impregnation), (b) Ag-Au/CNT-DP (deposition precipitation), (c) Ag-Au/CNT-MW (microwave synthesis).

EDS was employed to identify the elemental composition of the prepared samples. Several regions of about 1 mm^2 were chosen to measure the elemental distribution and the results for the metal deposited CNTs are shown in figure 2. All the metal functionalized samples show the presence of Ag and Au along with the peaks of C and O. No peaks for Ca, Fe and Co, used as catalysts during CNTs synthesis, were observed in any of the samples indicating the effectiveness of nitric acid treatment to remove all impurities from CNTs.

TEM

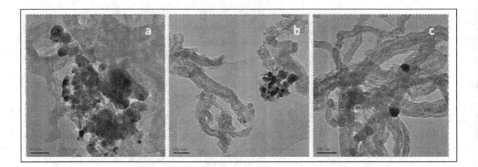

Figure 3. TEM images of Ag-Au loaded CNTs: (a) Ag-Au/CNT-WI (wet impregnation), (b) Ag-Au/CNT-DP (deposition precipitation), (c) Ag-Au/CNT-MW (microwave synthesis)

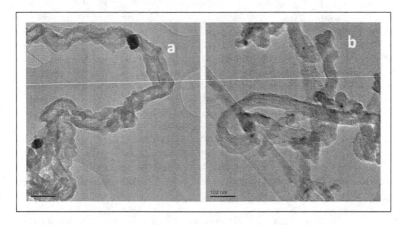

Figure 4. TEM images of (a) Ag/CNTs (b) Au/CNT

The TEM images (refer figure 3) show the significant amount metal nanoparticle aggregation in conventional methods of synthesis as compared to the microwave assisted synthesis. Microwave method is found to be advantageous in dispersing the metal nanoparticles more effectively than the other two methods. The particles appeared to be multicrystalline as seen from the images but the challenge remained as to differentiate which particle is silver and which one is gold. It is clearly observed that there is a particle size difference between the gold and the silver nanoparticles even at low magnification. In order to confirm the type of particles TEM analysis was done on CNTs decorated with individual metals. From the results (refer figure 4) it was clear that the bigger particles are of Ag nanoparticles. This result agrees with observations by Bin Xue et al. [31] who used a solid state preparation method for Ag and Au nanoparticles on CNTs. Most of nanoparticles observed are detached from the tubes perhaps during ultrasonication, showing that the adhesion of the nanoparticles to the CNTs is not strong probably due to the Vander Waals forces.

Raman

Figure 5 shows Raman spectra obtained for pure and metal loaded CNTs. It is observed that all the samples show a D-band at around 1340 cm^{-1} and the G-band at 1580 cm^{-1} which is typically expected for the CNTs. The intensity (I_G/I_D) ratios were calculated to be 0.991, 0.930, 0.903, and 0.906, respectively, for pure CNTs and composites from MW, DP and WI methods. The decrease in I_G/I_D ratios for the composite samples indicates the increase in disorder due to the presence of metal nanoparticles on the CNT surface. Comparatively samples prepared by conventional methods showed higher degree of disorder which may be due to the metal nanoparticle aggregations. These observations are in line with the TEM results.

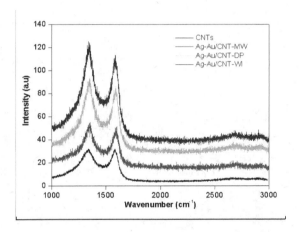

Figure 5. Raman spectra of CNT (pure), Ag-Au/CNT-WI (wet impregnation), Ag-Au/CNT-DP (deposition precipitation), Ag-Au/CNT-MW (microwave synthesis).

XRD

To further confirm the presence of Ag and Au particles, and check the crystalline phases, the prepared samples were analyzed by XRD and the diffractograms of various samples are presented in figure 6. All the samples show peaks of C (002) and C (100) phases with d_{002} values of 0.34 nm characteristic of CNTs. All the metal decorated samples show intense peaks corresponding to (111), (200), (220) and (311) for silver and gold nanoparticles at 37.96, 53.09, 77.38, respectively, which are known to have almost identical lattice constant [21]. The peaks at lower angles with 001,121 and 101 are probably due to the presence of gold chloride and it is important to note that these peaks are not there on the materials prepared by microwave method which shows that metals were reduced better.

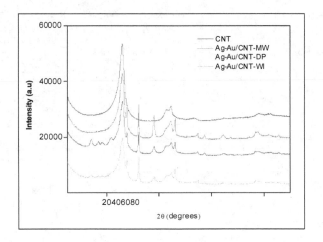

Figure 6. XRD patterns of CNT (pure), Ag-Au/CNT-WI (wet impregnation), Ag-Au/CNT-DP (deposition precipitation), Ag-Au/CNT-MW (microwave synthesis)

CONCLUSIONS

The synthesis of Ag-Au loaded CNTs composites was successfully done by the wet impregnation, deposition precipitation and microwave methods and the prepared samples were thoroughly characterized by various techniques. From the results it can be concluded that the microwave irradiation is advantageous over the conventional methods in getting much better dispersion of metal nanoparticles on CNTs surfaces. The characterization of the materials showed successful loading of the Ag and Au nanoparticles on the CNTs which would enhance the inherent properties of the CNTs support material. The mechanism of metal growth on the CNT surface is not conclusive yet which needs still more experimental evidences. The investigations on the optical and photocatalysis properties of Ag-Au decorated CNTs are underway.

REFERENCES

1. Cao, G., Nanostructures &Nanomaterials. synthesis, properties & applications. 2004: Imperial College press. 433.
2. Daenen, M., et al., The wondrous world of carbon nanotubes. a review of current carbon nanotubes technologies, 2003: p. 1-63.
3. Donaldson, K., et al., Carbon Nanotubes: A Review of Their Propertis in relation to Pulmonary Toxicology and Workplace safety. Toxicological science, 2006. **92**(1).
4. Khare, R. and S. Bose, Carbon Nanotubes Based composites- A review. Journal of Minerals & Materials Characterizations & Engineering, 2005. **4**(1): p. 31-46.
5. Li, C., E.T. Thostenson, and T.-W. Chou, sensors and actuators based on carbon nanotubes and their composites. Composites Science and Technology 2008(68).

6. Merkoci, A., Carbon Nanotubes in Analytical Chemisty. Microchimica Acta, 2005.

7. Tasis, D., et al., Chemistry of Carbon Nanotubes. American Chemical Society, 2006(106): p. 30.

8. Prabhuram, J., et al., Multiwalled Carbon Nanotube Supported PtRu for the Anode of Direct Methanol Fuel Cells. Journal of Physics and Chemistry, 2006. **110**: p. 7.

9. Nadagouda, M.N. and R.A. Varma, Noble Metal Decoration and alignment of Carbon Nanotubes in carboxymethyl cellulose. macromolecular jurnals. rapid communications 2008.

10. Motshekga, S., S.K. Pillai, and S.S. Ray, Conventional wet impregnation versus microwave-assisted synthesis of SnO2/CNT composites. journal of Nanopart Res, 2010: p. 7.

11. Mialinger, K.A., et al., Microwave frequency effect on synthesis of cyptomelane-type manganese oxide and catalytic activity of cryptomelane precursor. journal of catalysis, 2006. **239**: p. 8.

12. Li, X., et al., Microwave poly synthesis of Pt/Cnts catalysts: Effect of PH on particle size and electro calalytic activety for methanol electrooxidation. Carbon, 2005. **43**: p. 6.

13. Salvetat, J.P., J.M. Bonard, and N.H. Thompson, mechanical properties of carbon nanotubes. Applied physics A, 1999(69).

14. Mhlanga, S., et al., the effective synthesis parameters on the catalytic synthesis of multiwalled carbon nanotubes using Fe-Co/ CaCO$_3$. South African Journal of chemistry, 2009. **62**(67): p. 25.

15. Qiang, Z., Z. Mengqiang, and H. Jiaqi, selective synthesis of single/double/ multi-walled carbon nanotubes on Mgo-Supported Fe Catalyst. Chinese Journal of catalysis, 2008. **29**(11): p. 6.

16. Lin, K.-N. and T.-y. yang, A novel method of supporting gold nanoparticles on MWCNTs: Synchroton X-ray reduction. china particualogy, 2007. **5**: p. 4.

17. Tanaka, N., et al., Photochemical deposition of Ag nanoparticles on multiwalled carbon nanotubes. Carbon, 2009. **47**: p. 8.

18. Tello, A., et al., the synthesis of hybrid nanostructures of gold nanoparticles and carbon nanotubes and their transformation to solid carbon nanorods. carbon, 2008. **46**: p. 5.

19. capek, I., Dispersions, novel nanomaterial sensor and nanocongugates based on carbon nanotubes. aAdvances in colloid and interface science, 2009. **150**: p. 26.

20. Hiramatsu, M. and M. Hori, Perparation of Dispersed Platinum Nanoparticles on a Carbon Nanostructured Surface using Supercritical Fluid Chemical Deposition. Materials, 2010. **3**: p. 12.

21. Pal, A., S. Shah, and S. Devi, Preparation of silver-gold and gold-silver bimetalic nanoparticles in w/o microemulsion containing TritonX100. Science Direct, 2007. **302**: p. 4.

22. Pan, B., et al., Attachment of gold nanoparticles on multi-walled carbon nanotubes. Nanoscience, 2006. **11**(2): p. 95-101.

23. Tong, H., H.-L. Li, and X.-G. Zhang, Ultrasonic synthesis of highly dispersed Pt nanoparticles supported on MWCNTs and their electrocatalytic activity towards methanol oxidation. Carbon, 2007. **45**: p. 8.

24. Xu, P., et al., A facile stratergy for covalent binding of nanoparticles onto carbon nanotubes. Applied Surface Science, 2008. **254**.

25. Zhang, M., L. Su, and L. Mao, surfactant functionalization of carbon nanotubes (CNT) for layer-by-layer assembling of CNT multi-layer films fabrication of gold nanoparticles/ CNT nanomayer. Carbon, 2006. **44**: p. 7.

26. Li, Z., et al., Methan sensor based on nanocmposite of Palladium/ multi-walled carbon nanotubes grafted with 1,6- hexanediamine. Sensors and Actuators B: chemical, 2009. **139**: p. 6.

27.   Zhang, W., et al., Microwave-assisted synthesis of Pt/CNT nanocomposite electrocatalysts for PEM fuel cells. The royal society of chemistry 2009(2): p. 282.

THE COMPARISON IN THE EFFICIENCY IN NITROGEN DOPED CARBON NANOTUBES AND UNDOPED CARBON NANOTUBES IN THE ANCHORING OF SILVER NANOPARTICLES

K Mphahlele[1,2], S. Sinha Ray[1], M.S Onyango[2] and S. D Mhlanga[3]
[1]DST/CSIR Nanotech. Innovation Centre, NCNSM, PO Box 395, Pretoria 0001, South Africa.
[2]Faculty of Engineering and the Built Environment, Tshwane University of Technology, Private Bag X680, Pretoria 0001, South Africa.
[3]School of Chemistry, University of the Witwatersrand, P.O. Wits 2050, Johannesburg, South Africa.
*Corresponding author. E-mail: KMphahlele@csir.co.za

ABSTRACT

Nitrogen doped carbon nanotubes (N-CNTs) provide unique structure, controlled electrical properties and strong interactions with deposited nanoparticles due to the presence of nitrogen. The aim of this communication is to present the preliminary results obtained after investigating the efficiency of N-CNTs and undoped CNTs in the anchoring of silver (Ag) nanoparticles. N-CNTs synthesized in a tubular quartz reactor that was inserted in a horizontal furnace were used. Microwave assisted polyol synthesis and wet impregnation method was used to successfully load Ag nanoparticles on both N-CNTs and undoped CNTs. The samples have been characterized by transmission electron microscopy (TEM), scanning electron microscopy (SEM), inductively coupled plasma optical emission spectrometry (ICP-OES), X-Ray Diffraction (XRD), BET, and Raman Spectroscopy. N-CNTs provide a good support for Ag nanoparticles despite a lower surface area and less surface defects when compared to undoped CNTs. The metal particles were uniformly dispersed on the surface of the N-CNTs. The results suggested that the N-CNTs have strong interactions with the metal nanoparticles which may prevent agglomeration / sintering of the metal nanoparticles during use in water purification.

INTRODUCTION

Over the last decade, carbon nanotubes (CNTs) have been the subject of extensive research due to their outstanding mechanical and electronic properties. It has been shown that CNTs can behave as metals or semiconductors depending on their diameter and chirality. Although CNTs can be produced using a number of methods, there is still no control over nanotube morphology and hence their electronic properties.[1] However, theoretical[2] and experimental studies[3] revealed that it is possible to tune the electronic properties of the nanotubes by incorporating hetero atoms within the carbon lattice.[4] The most frequently used dopants are boron and nitrogen because their atom size is similar to that of carbon[5] and because they serve as p- or n-type dopants, respectively.[4] The successful incorporation of nitrogen[6] or boron[5] atoms within the graphitic carbon cylinders strongly depends on the choice of precursor, catalyst, reaction temperature, reaction time, gas flow rate and pressure.[7] Furthermore, the doping of nanotubes with hetero atoms also changes the nanotube structure, chemical reactivity and mechanical stability. Doping of CNTs could lead to a significant alteration of their intrinsic physical and chemical properties and in turn, their catalytic activity and selectivity as well. The chemically modified CNTs surfaces could be further used as a platform for various applications such as N-CNT film electrodes. Due to the small difference in atomic radii of carbon compared to nitrogen and boron atoms, they are the two most widely used as substitution elements inside the carbon matrix. The electronic modification induced by the foreign elements doping, increases their sensitivity towards different gaseous molecules compared to the pristine CNTs.[8]

Recently, a simple and rapid synthesis method for metal catalysts using microwave irradiation energy has been suggested. The advantage of the microwave irradiation is that it transfers heat to the substance uniformly through the microwave-transparent reaction container (i.e. glass or plastic), leading to a more homogeneous nucleation and shorter crystallization time compared to the conventional heating during which unavoidable temperature gradients occur, in particular when large volumes of solutions are used. This gradient of temperature may adversely affect the particle size distribution and yield.[9] In the present paper, the anchoring of silver (Ag) particles on the N-CNTs and undoped CNTs was investigated and compared using microwave polyol method and conventional wet impregnation.

EXPERIMENTAL

Synthesis of carbonaceous material

N-CNTs were synthesized by the decomposition of a ferrocene-triethylamine-toluene solution mixture in a tubular quartz reactor that was inserted in a horizontal furnace. The solution, placed in a syringe, was injected into the tube using an electronically controlled injector at high temperatures. The ferrocene was used as the catalyst while triethylamine was the source for nitrogen.

Undoped CNTs were synthesized by the decomposition of acetylene (C_2H_2) using $CaCO_3$ supported Fe-Co catalysts in a tubular quartz reactor that was inserted in a horizontal furnace. The gas flow rates were accurately controlled as desired using mass flow controllers.

Loading of silver particles

Microwave assisted polyol method

Undoped CNTs and N-CNTs were dispersed in ethylene glycol in separate beakers and were sonicated at room temperature, two separate Ag were also poured in ethylene glycol and also sonicated at room temperature. The Ag solution was poured into the CNTs solution. The solution was put in the Multiwave 3000 microwave tubes at 473 K for 15 min, the solution was filtered and acetone was used to clean the CNTs. The precipitate was dried overnight in oven.

Conventional wet impregnation method

N-CNTs and undoped CNTs were dispersed in distilled water by sonication, in separate beakers. Then silver nitrate solution was added slowly with vigorous stirring. The solution was stirred at room temperature for at least 20 h and separated by four centrifugation cycles with each cycle at a rate of 9000 rpm with a 25 ml plastic centrifuging tube. In each case 50 ml of water was added to the plastic tube to aid dispersion of the collected solid. The precipitate was then washed several times with distilled water and dried overnight in an oven.

Leaching test

5 mg of Ag/N-CNTs and Ag/undoped CNTs were each sonicated in 25 mL of distilled water (in a beaker) for different times such as 10, 30 and 60 min. After sonication, the materials were then separated from the solvent (water) using Munktell filter paper and the resultant water samples (filtrates) were measured for the presence of silver using Inductively Coupled Plasma Emission Spectrometry (ICP-OES).

5 mg of Ag/N-CNTs and Ag/undoped CNTs were each dispersed in 25 mL of distilled water (in a beaker) for different times (30, 60 min and overnight). The materials were then separated from the solvent (water) using Munktell filter paper and the resultant water samples (filtrates) were measured for the presence of silver using Inductively Coupled Plasma Emission Spectrometry (ICP-OES).

Characterization

Characterization of the carbonaceous materials was performed by transmission electron microscopy (TEM) and scanning electron microscope (SEM). The water samples were characterized using Vista-Pro CCD simultaneous-Varian ICP-OES. TEM analysis was performed on a JEOL JEM-2100 TEM at 200 kV and at varying magnifications. The sample for TEM was prepared by the soniction of about 5.0 mg of the CNTs in ethanol and several drops were dropped onto a TEM grid. X'Pert Pro PANalytical x-ray diffraction (XRD) was used to observe x-ray diffraction patterns of N-doped and undoped CNTs decorated with Ag nanoparticles. Micromeritics VacPrep 061 Sample Degas System Branuer-Emmett-Teller (BET) was used to investigate the surface area and porosity of the samples. Raman Horiba Jobin Yvon HR800 UV Spectroscopy was used.

RESULTS AND DISCUSSION

Scanning electron microscope analysis

The morphology of the undoped CNTs and N-CNTs before and after loading with metals was studied by TEM and SEM. Results are summarised in Figures 1 and 2. All the CNTs used in this study were multi-walled. SEM and TEM analysis revealed that the N-CNTs were more aligned in bundles and where characterized by compartments in the wall of the CNTs (*bamboo-like* structure) with outer diameters between 30-50 nm (refer Figure 1a). The undoped CNTs did not have the bamboo-like structure observed in the case of N-CNTs. Rather, they were randomly oriented in a 'spaghetti-like' fashion, also with a narrow size range between 45 and 55 nm (refer Figure 1b).

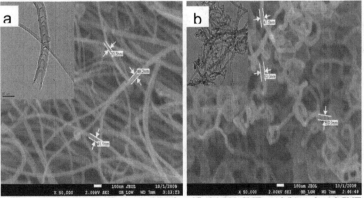

Figure 1. SEM and TEM images (inset) of purified (a) N-CNTs and (b) undoped CNTs.

SEM images of Ag particles on N-CNTs and undoped CNTs was observed in Figure 2 in which was loaded using microwave polyol and wet impregnation methods. The Ag nanoparticles (white spots) were uniformly dispersed on the surface of the N-CNTs that were loaded using microwave polyol method (refer Figure 2a). SEM images of Ag/undoped CNTs also showed Ag nanoparticles (white spots) ,but some aggregation on some parts of the sample (refer Figures 2b and 2d). However a poor dispersion of the Ag nanoparticles on the CNTs was observed when wet impreganation was used to load the Ag particles on both N-CNTs and undoped CNTs surfaces (refer Figures 2c and 2d).

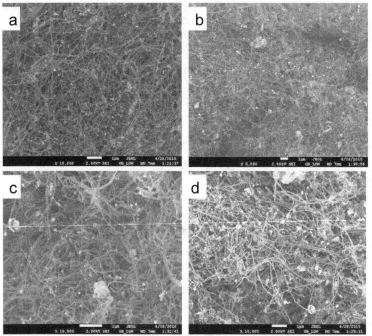

Figure 2. SEM images of microwave polyol synthesis (a and b) and wet impregnation (c and d): (a) Ag/N-CNTs, (b) Ag/undoped CNTs, (c) Ag/N-CNTs and (c) Ag/ undoped CNTs.

Transmission electron microscopy analysis

TEM images of Ag/N-CNTs is shown in Figures 2a and 2b in which Ag particles were loaded using microwave polyol and wet impregnation methods respectively. The Ag nanoparticles (dark spots) had diameters in the range 3 – 5 nm and were uniformly dispersed on the surface of the N-CNTs (refer Figure 2a). TEM images of Ag/undoped CNTs is shown in Figure 2b and 2d in which Ag particles were loaded using microwave polyol and wet impregnation methods respectively. The Ag nanoparticles (dark spots) have diameters in the range 3 – 5 nm. Microwave polyol method proved to be more superior to the wet impregnation method (refer Figures 3c and 3d), as better dispersion of Ag particles was observed on both N-CNTs and undoped CNTs that were decorated with Ag particles which was loaded using microwave polyol method (refer Figures 3a and 3b).

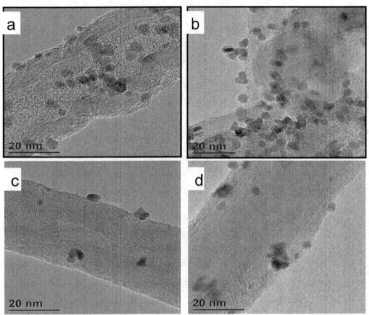

Figure 3. TEM images of microwave polyol synthesis (a and b) and wet impregnation (c and d): (a) Ag/N-CNTs, (b) Ag/undoped CNTs, (c) Ag/N-CNTs and (c) Ag/ undoped CNTs.

X-ray mappings analysis

Several regions of about 1 mm^2 were selected to measure the elemental distribution using EDS and the results are shown in Figure 4. The X-ray mappings which connected to SEM of the selected regions indicate the presence of high densities of the metal particles (purple = Ag) on the CNTs. It could be observed from the X-ray mappings that the dispersion of the Ag particles was more uniform on the N-CNTs than on the undoped CNTs for all material. The particles were more aggregated on the undoped CNTs. The Ag particles that were loaded using microwave method showed a more homogeneous dispersion on the surface of the CNTs (refer Figures 4a and 4b) compared to those loaded with wet impregnation method (refer Figures 4c and 4d). These observations shows that wet impregnation method often lack good control of particle size and morphology as it is well known that catalytic activity of metal particles is strongly depended on the particle size and distribution. This agrees with the observation from HRTEM analysis, which revealed that the N-CNTs have less defects on the surface than the undoped CNTs.

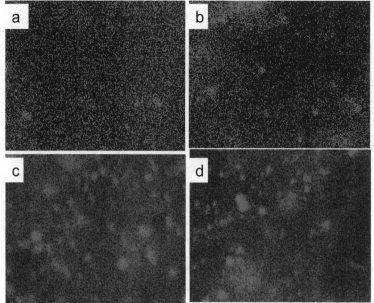

Figure 4. X-ray mapping images of microwave polyol synthesis (a and b) and wet impregnation (c and d): (a) Ag/N-CNTs, (b) Ag/undoped CNTs, (c) Ag/N-CNTs and (c) Ag/ undoped CNTs.

X-ray diffraction analysis

Parts (a) and (b) of Figure 5a and 5b respectively, shows the XRD pattern for the undoped CNTs, N-CNTs and the CNTs after Ag nanoparticle decoration. It is possible to observe reflections arising from the Ag particles. The CNTs revealed reflections corresponding to the C(100) planes of crystalline graphite-like materials, whereas the Ag/undoped CNTs and Ag/N-CNTs showed those corresponding to four main crystallographic planes, namely, Ag(111), Ag(200), Ag(220), Ag(311) and Ag(222). These results confirm that Ag nanoparticle where loaded using wet impregnation and polyol microwave.

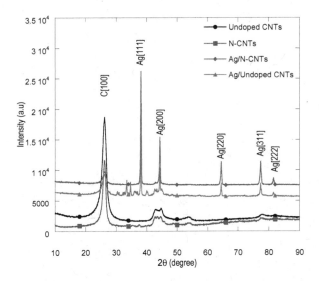

Figure 5a: XRD pattern before and after Ag decoration using microwave polyol method.

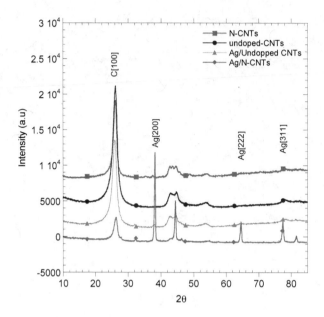

Figure 5b: XRD pattern before and after Ag decoration using conventional wet impregnation method.

Raman spectroscopy study

The quality of the bulk samples of the CNTs was studied using Raman spectroscopy and results are presented in Figure 6. Two maxima appeared in both spectra, near 1355 and 1585 cm^{-1}. The I_G/I_D ratio of N-doped and undoped CNTs of 0.919 and 1.06 respectively was observed. The ratio between the D- and G- bands suggest that the N-doped CNTs have less structural defects than the undoped CNTs. Indeed, characterization of the CNTs using HRTEM supports this explanation (see Figure 7). The graphite layers of N-CNTs are smoother, more ordered and have fewer defects (see Figure 7a). On the contrary, the undoped CNTs possessed a high amount of defects on the wall structure as shown in Figure 6b. The observed difference in the structure of the CNTs is due to the different carbon sources and experimental conditions used in synthesis in CVD.

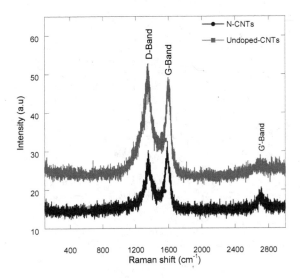

Figure 6: Raman Spectroscopy of N-CNTs and undoped CNTs

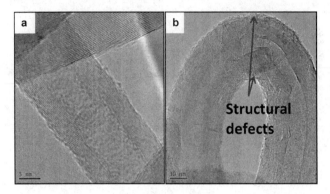

Figure 7: shows the HRTEM of (a) N-CNTs and (b) undoped CNTs

Branuer-Emmett-Teller (BET) surface area analysis

Table 1 shows the BET surface areas, pore volumes and average pore sizes of the raw and oxidized undoped and N-CNTs obtained using a Micromeritics TriStar II Surface Area and Porosity Analyzer. The results indicate that the oxidized CNTs have higher surface areas because most of the reagent residues have been removed. The undoped CNTs have higher surface areas and pore volumes than the N-CNTs. This is attributed to the higher surface defects and open tubes (no compartments) of the undoped CNTs.

Table 1: BET surface area and porosity analysis of the undoped and N-CNTs.

Sample name	BET surface area (m^2/g)	Pore volume (cm^3/g)
Raw CNTs	75	0.24
Oxidized CNTs	98	0.31
Raw N-CNTs	25	0.10
Oxidized N-CNTs	52	0.17

Leaching analysis

The results presented in Table 2 for the Ag/undoped CNTs samples suggest that the leaching of Ag nanoparticles increases as the sonication time increases. For the Ag/N-CNTs sample, the results suggest that there was no leaching of Ag nanoparticles or the leached Ag nanoparticles was too small to detect. Such an observation confirms that doped sites within CNTs significantly modify the chemical reactivity. For the N-CNTs sample, the results showed that there was no leaching of Ag nanoparticles for time 10, 30 and 60 min or no Ag was detected.

Table 3 shows that the concentration of silver nanoparticles (mg/L) remaining in water after suspending the CNTs in water was below the detection limit of the ICP-OES. This suggests that there was no leaching of silver nanoparticles or the concentration of leached silver nanoparticles was too small to be detected.

Table 2 Shows that the concentration of Ag molecules (mg/L) remaining in water after sonication the samples in water

Time (minutes)	Concentration of silver (mg/L)	
	Ag/undoped CNTs	Ag/N-CNTs
10	0.04	<0.03
30	0.07	<0.03
60	0.08	<0.03

Table 3: Shows that the concentration of silver molecules (mg/L) remaining in water after suspending the samples in water

Time (minutes)	Concentration of silver (mg/L)	
	Ag/undoped CNTs	Ag/N-CNTs
30	<0.03	<0.03
60	<0.03	<0.03
overnight	<0.03	<0.03

CONCLUSIONS

N-CNTs provide a good support for Ag nanoparticles despite a lower surface area and less surface defects when compared to undoped CNTs. The metal particles were uniformly dispersed on the surface of the N-CNTs. The results suggest that the N-CNTs have strong interactions with the Ag nanoparticles which may prevent agglomeration / sintering of the Ag nanoparticles during use in water purification and catalytic reactions. The results obtained show that during sonication, N-CNTs were more efficient in the anchoring of Ag nanoparticles. This is due to the nitrogenated sites that are chemically active. These active sites do not only result to a strong Ag-CNT interaction but also allow a uniform deposition of Ag along the nanotubes. On the other hand, sonication of the undoped CNTs resulted to increase in the leaching of the silver as the sonication time was increased. This suggests that the silver may leach back into solution under harsh/vigorous naturally occurring conditions. The N-CNTs have however shown to be more efficient materials for binding silver molecules even under vigorous conditions when compared to undoped CNTs. Under normal stagnant conditions no leaching was found for all the CNTs. Ag/N-CNTs and Ag/undoped prepared by microwave polyol method exhibited better dispersion of particles on the support material.

ACKNOWLEDGMENT

The authors wish to thank DST and CSIR, South Africa for the financial support.

REFERENCES

[1]A. Koos, M. Dowling, K. Jurkschat, A. Crossley and N. Grobet, Effect of the Experimental Parameter on the Structure of Nitrogen-Doped Carbon Nanotubes Produced By Aerosol Chemical Vapour Deposition, *Carbon*, **47**, 30-37 (2009).
[2]S. Latil, S. Roche, D. Mayou and J. Charlier, Mesoscopic Transport in Chemically Doped Nanotubes, *Physical Review Letters*, **92**, 256805 (2004).
[3]X. Wu, Y. Tao, C. Mao, L. Wen and J. Zhu, Synthesis of Nitrogen-Doped-Horn-Shaped Carbon Nanotubes by Reduction of Pentachloropyridine with Metallic Sodium, *Carbon*, **45**, 2253-2259 (2007).
[4]M. Terrones, A. Jorio, M. Endo, A. Rao, Y.Kim, T. Hayashi, H. Terrones, J. Charlier, G. Dresselhaus and M. Dresselhaus, New Direction in Nanotube Science, *Materials Today*, **7**, 30-45 (2004).
[5]O. Stephan, P. Ajayan, C. Colliex, P. Relich, J Lambert, P. Bernier and P Lefin, Doping Graphitic and Carbon Nanotube Structures with Boron and Nitrogen, *Science*, **266**, 1683-1685 (1994).
[6]S Kim, J. Lee, C. Na, J. Park, K. Seo, and B. Kim, N-doped Double-walled Carbon Nanotubes Synthesized by Chemical Vapor Deposition. *Chemical Physics Letters*, **413**, 300-305 (2005).
[7]C. Tang, Y. Bando, D.D. Golberg and F. Xu, Structure and Nitrogen Incorporation of Carbon Nanotubes Synthesized by Catalytic Pyrolysis of Dimethylformamide, *Carbon*, **42**, 2625-2633 (2004).
[8]K. Chizari, I. Janowska, M. Houlle, I. Florea, O. Ersen, T. Romero, P. Bernhardt, M. Ledoux and C. Pham-Huu, Tuning of Nitrogen-Doped Carbon Nanotubes as Catalyst Support for Liquid-Phase reaction, *Applied Catalysis A: General*, **380**, 72-80 (2010).
[9]M. Sakthivel, A. Schlange, U. Kunz and T. Turek, Microwave Assisted Synthesis of Surfactant Stabilized Platinum/Carbon Nanotube Electrocatalysts for Direct Methanol Fuel Cell Applications, *Journal of Power Sources*, **195**, 7083-7089 (2010).

Synthesis, Functionalization, and Processing of Nanostructured Material

SYNTHESIS OF SUBMICRON-SIZE CaB$_6$ POWDERS USING VARIOUS BORON SOURCES

A.Akkoyunlu*[a], R. Koc[a], J. Mawdsley[b] and D. Carter[b]

[a] Department of Mechanical Engineering and Energy Processes, Southern Illinois University Carbondale, IL 62901

[b] Argonne National Laboratory
9700 S. Cass Avenue
Argonne, IL 60439

ABSTRACT

This paper deals with the synthesis of submicron calcium hexaboride (CaB$_6$) powders from different mixing methods. This research makes use of various boron sources to compare the mixing techniques and properties of final products. X-ray diffraction (XRD), transmission electron microscopy (TEM), and (BET) were utilized to investigate the properties of resulting CaB$_6$ powders. XRD and TEM data for partially coated precursors containing B$_4$C and CaCO$_3$ demonstrate the superiority of the partial coating in the formation of CaB$_6$. These precursors produced about 150 nm size CaB$_6$ particles at 1550°C for 4 hrs in flowing argon.

1. INTRODUCTION

Calcium hexaboride (CaB$_6$) has very unique combination of properties such as high chemical stability, high melting point (2235°C), tolerable hardness and high electrical conductivity. It has a low density (2.45 g/cm^3) and the crystal structure is cubic lattice with boron octahedra located at each of the eight corners and one calcium rare-earth atom at the center of the cube [1]. Because CaB$_6$ is utilized as wear resistant material, abrasive and surface protection material, recently it is developed as a filler material in bipolar plates of proton exchange membrane (PEM) fuel cells [2]. This creates a gradually ascending interest as a part of advanced technology.

As the attention increases, the number of investigations about CaB$_6$ also has increased recently. But there is still very limited information in literature about the synthesis of CaB$_6$ powder [3]. For this reason, this paper describes one of the synthesis methods of CaB$_6$ which is prepared by using boric acid (H$_3$BO$_3$), boron carbide (B$_4$C), calcium carbonate (CaCO$_3$), and carbon (C) powders. Moreover, the XRD, TEM and BET results of the CaB$_6$ powder which is reacted at different temperatures in the tube furnace are reported. The aim of this study is to form highly efficient and economical CaB$_6$ powder which can be used in fuel cells as a filler material for manufacturing bipolar plate composites.

2. EXPERIMENTAL PROCEDURES

2.1. Starting Materials

In this research, a number of methods were used to produce high quality low cost calcium hexaboride (CaB$_6$) powders. These methods are based on carbothermal reduction reactions as explained in Dr. Koc's patents developed for the production of SiC (U.S. patent:#5324494)[4], and TiC, TiN (U.S. patent: #5417952)[5]. The following starting materials were chosen for the research: C (carbon black) *Cabot Corporation – Monarch 880*, CaCO$_3$ (calcium carbonate) *Specialty Minerals – Vicality Extra Light*, B$_4$C (boron carbide), and H$_3$BO$_3$ (boric acid) *Spectrum BO110*.

127

2.2. Preparation of Precursors

The preparation of precursor is separated into two different groups. One is mixed type precursors. And the other one is coated type precursors prepared using a propriety method which is subject to patent application.

2.2.1. Mixed Type Precursors

In this study, two different reactions were performed and in each reaction three types of powders were used as starting materials; (1) carbon (C), calcium carbonate and boron carbide, (2) carbon, boric acid and calcium carbonate. Mixtures in (1) and (2) are reacted into their final products in tube furnace at 1473°C and 1375°C respectively as displayed in the following equations:

$$2CaCO_3 + 3B_4C + C \longrightarrow 2CaB_6 + 6CO_{(g)} \qquad ---- (1)$$

$$CaCO_3 + 6H_3BO_3 + 11C \longrightarrow CaB_6 + 9H_2O + 12CO_{(g)} \qquad ---- (2)$$

In all of the mixture, the necessary materials were prepared using the stoichiometric calculations for the reactants. First, C, $CaCO_3$ and B_4C were mixed with weight percents of 3.178, 52.965, and 43.857 respectively. Then, C, H_3BO_3 and $CaCO_3$ were mixed with weight percents of 21.903, 61.504, and 16.593 respectively for the second precursor.

The stoichiometrically pre-weighed powders were ball milled (Model Spex Turbula 8000) in a plastic bottle for 30 minutes. After milling, the mixture was poured into a graphite crucible and they were weighed before the reaction. The crucible was placed inside a tube furnace while the high flow rate of argon gas was blowing through the tube. The heating rate of the furnace was 5°C min⁻¹. The holding time of the materials was 4 hours at each reaction temperature. The argon gas was flowed continuously at high rate during the reaction. The lid of the furnace was opened when the inner temperature was less than 100°C. The crucible was removed from the furnace and weighed again to check the weight loss. The samples were subjected to X-ray (Model Rigaku Mini Flex II Dekstop X-ray difffractometer) for phase analysis. To determine the particle size and obtain information on particle morphology the samples investigated under transmission electron microscopy (TEM) (Model Hitachi H7650-II).

2.2.2. Coated Type Precursors

The coated precursor was prepared by cracking C_3H_6 gas at 550°C, with pyrolytic carbon, coating the $CaCO_3$. Each oxide particle was uniformly coated by similar amount of porous amorphous carbon, as this process is surface area activated. Two types of coated precursors were prepared. One is called 10 cycle and the other one is labeled as 30 cycle. In each cycle, the $CaCO_3$ was coated for 20 minutes.

The micrographs of $CaCO_3$, the coated precursor – 10 cycle and 30 cycle carbon coated precursors are shown in Figure 5. Mixing was performed using appropriate amounts of $CaCO_3$ – C or $CaCO_3$ and carbon black.

2.3. Characterization of Samples

The methods employed in this research were the determination of the surface area (BET), X-ray diffraction (XRD) and, transmission electron microscope (TEM).

2.3.1. X-ray Diffraction (XRD)

The analytic applications of X-ray diffraction were first developed by Laue and his students in the early 20th century. From that day, XRD is still advancing rapidly in the several types of problems it faces with [6]. Basically, when X-ray hits a crystalline solid, constructive interference effect of elastically scattering X-ray results.

The phase analysis and characterization of the starting materials and, reaction products were employed using X-ray diffractometer (Model Rigaku Mini Flex II Dekstop X-ray difffractometer, with a scan rate of 2°C min^{-1}) to ensure the phases formed properly. The peaks broaden when the angles are high and long wavelengths occur when the protons have low level of energy.

2.3.2. Transmission Electron Microscope (TEM)

Electron microscopy is specified as a specialized field of science that performs the electron microscope as a tool. By magnifying the samples of products thousands times, the powder samples' morphology were observed and analyzed via transmission electron microscopy (TEM) (Model Hitachi H7650-II). Before using TEM, a specific preparation was carried out. Firstly, each reaction product sample was ground; secondly appropriate amount of alcohol was poured to each sample roughly, and then, the solutions were cleaned ultrasonically for ten minutes, finally they were kept to dry for ten hours. Basically, the aim of this preparation was to analysis the samples easily and to get more precise results from TEM.

2.3.3. The Determination of Surface Area (BET)

The surface area of the starting materials and reacted powders were measured using BET surface area analyzer (Gemini 236, Micromeritics, Norcross, GA). BET process was employed on 12 different points of the each sample surface. Then, all samples were degassed for 4 hours at 150°C in flowing nitrogen atmosphere. During BET process helium gas was also used to analyze the surface area

3. RESULTS AND DISCUSSION

3.1. XRD Results

The reactions (1) and (2) were reacted at several temperatures for 4 hrs in flowing argon. These starting materials started to react at about 700°C. Then, transition phases Ca$_3$B$_2$O$_6$, CaO occurred at 1000°C. Fig. 1 and 2, show all the transition phases, single phases formed with traditional mixing method. When the temperature goes up, CaB$_6$ product peaks appeared with having weak intensity. Single phase barely observed in 1550°C after the reaction (1) performed. But after reaction (2) was completed, single phases started forming at 1450°C, although they were not very strong peaks.

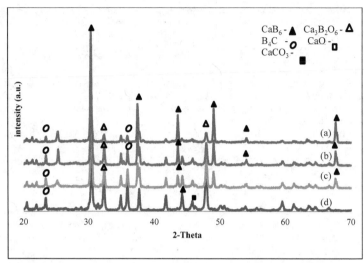

Figure 1. X-ray diffraction patterns of the product of reaction (1) reacted at several temperatures ((a) 1550°C (b) 1450°C, (c) 1350°C, and (d) 1250°C).

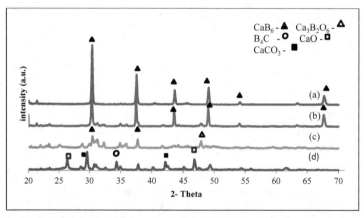

Figure 2. XRD patterns of the product of reaction (2) reacted at several temperatures ((a) 1550°C, (b) 1450°C, (c) 1350°C, and (d) 1250°C).

XRD shows that beside the targeted CaB_6, other phases such as $Ca_3B_2O_6$, CaO are present in the final product. Because these results did not satisfy the objective designated at the beginning, it was decided to increase the carbon amount. One of the ways to do this was to coat the powders with carbon.

Then, 10 cycle and 30 cycle carbon coated precursors were included for both of the reaction (1) and (2) and those reactions were reacted at 1550°C for 4 hrs in tube furnace under flowing argon.

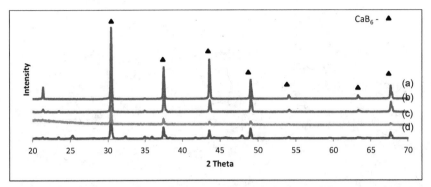

Figure 3. XRD results of commercially available CaB₆ powder and three different reacted precursors at 1550°C for 4hrs based on reaction 1, (a) Commercial CaB₆, (b) B₄C-30Cycle, (c) B₄C-10Cycle, (d) B₄C traditionally mixed.

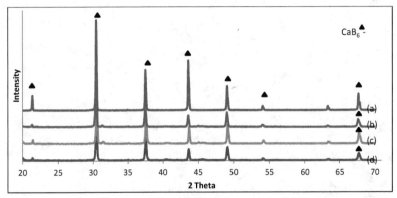

Figure 4. XRD results of commercially available CaB₆ powder and three different reacted precursors at 1550°C for 4hrs based on reaction 2. (a) Commercial CaB₆, (b) H₃BO₃-30Cycle, (c) H₃BO₃-10Cycle, (d) H₃BO₃ traditionally mixed.

In the XRD results of the carbon coated precursors, XRD of commercially available CaB₆ powder was also included to make a comparison. And, it was obtained that the products from the reactions performed at 1550°C for 4hrs were single phases and the intensity of the product peaks as strong as commercial one. Thus, it was observed from TEM micrographs in Fig. 5., the products size is smaller than the commercial one. These results are also consistent with the objective of production of submicron powders.

These results clearly show that the formation of single phase CaB$_6$ was possible for all the precursors, including traditionally mixed starting materials when the boron source was boric acid. Boric acid is more economical as compared to the boron carbide as a boron source in the process.

3.2. TEM Micrographs

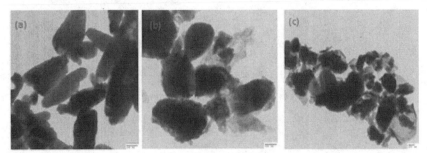

Figure 5. TEM micrographs of (a) commercial CaCO$_3$, (b) 10-cycle carbon coated CaCO$_3$ and (c) 30-cycle carbon coated CaCO$_3$. (Scale bar shows 100 nm.)

Fig. 5, shows the TEM micrographs of starting powder CaCO$_3$ (Vicality Extra Light-Specialty Minerals) and two different precursors. Figure 5b and 5c are TEM of carbon coated CaCO$_3$ showing the quality of the precursors, mainly uniformity of pyrolitic carbon layer (light gray color) on the particles of CaCO$_3$.

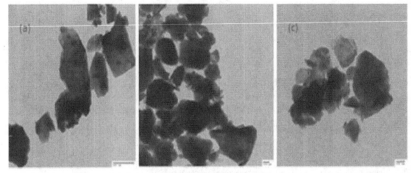

Figure 6. TEM micrographs of resulting CaB$_6$ powder from reaction 1 and commercially available CaB$_6$ powders for comparison. (a) Commercial CaB$_6$ (b) CaB$_6$ from 10-cycle carbon coated precursor (B$_4$C) reacted at 1550°C for 4 hrs, and (c) CaB$_6$ from 30-cycle carbon coated precursor (B$_4$C) reacted at 1550°C for 4 hrs. (Scale bar shows 500 nm in fig. 6a and 100 nm in fig. 6b,6c.)

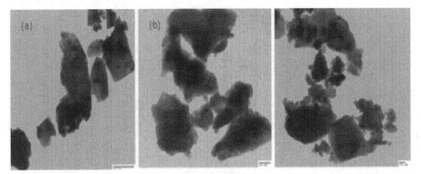

Figure 7. TEM micrographs of resulting CaB$_6$ powder from reaction 2 and commercially available CaB$_6$ powders for comparison. (a) Commercial CaB$_6$ (b) CaB$_6$ from 10-cycle carbon coated precursor (H$_3$BO$_3$) reacted at 1550°C for 4 hrs, and (c) CaB$_6$ from 30-cycle carbon coated precursor (H$_3$BO$_3$) reacted at 1550°C for 4 hrs. (Scale bar shows 500 nm in fig. 7a and 100 nm in fig. 7b,7c.)

TEM micrographs show that the particles of CaB$_6$ synthesized using the new method are submicron-size, uniform and have no agglomeration. There is also a similarity between the starting powder CaCO$_3$ and formed CaB$_6$ with respect to size and shape which are seen from the TEM micrographs. This is a proof that the starting materials have a direct effect on the products. These results indicate that the particle sizes of the resulting products were much finer because boric acid being the boron source. It was also interpreted from the images that when the number of the coating cycle increased, size of the particle increased too. Coated carbon can be seen clearly around the CaB$_6$ particles.

3.3. BET Results

Furthermore, the surface areas of products were measured by BET method. Surface area changes observed with varying reaction temperatures. The amount of the sample used in test tubes was between 0.5 and 1.1 grams. Before measuring the surface areas, all samples were degassed to prevent the faulty results caused from the impurities.

Table I. BET results from reaction 1

Temperatures (°C)	Sample Weight (g)	Surface area (m^2/g)
1250	0.5625	16.6165
1350	0.8492	11.9123
1450	1.0623	3.3347
1550	0.8796	1.9522

Table II. BET results from reaction 2

Temperatures (°C)	Sample Weight (g)	Surface area (m^2/g)
1250	0.6120	2.4750
1350	0.9107	2.1732
1450	0.7785	2.0037
1550	0.8240	1.6603

From 1250°C to 1550°C, a steady reduction of surface area (m^2/g) occurred for both sets of samples as tabulated in Table 1 and II. Carbon coating process and increasing the heating temperatures caused to increase the size of the powder particles. As the particle size of the powder increases, the surface area of the powder decrease significantly.

4. CONCLUSION

Single phase product was obtained with the particle size of 200-400 nm. The effects of heating temperature and carbon coating process during the production of CaB$_6$ powder showed that the optimum heating temperature is 1550°C for reaction (1) and 1450°C for reaction (2) in traditional method and, the optimum heating temperature is 1350°C for reaction (1) and 1250°C for reaction (2) in carbon coating method. The morphology proved that CaB$_6$ formed from both of the reactions has a finer particle size than commercial one. Because the morphology of the coated starting material CaCO$_3$ and formed product CaB$_6$ have similarities with respect to size and shape, it can be mentioned that starting materials affects the formation of product directly. As the temperature is increased, particles became less agglomerated, shapes became more spherical, and the size also decreased [7]. CaB$_6$ powder which was synthesized by carbon coated method has a lower formation temperature than the product synthesized by traditionally mixing method. Powder produced using the coated precursor/traditionally mixed method have the following characteristics; fine size, pure and single phase, spherical, narrow size distribution, non-agglomerated.

ACKNOWLEDGEMENT

This Research has been sponsored by the U.S. Department of Energy through Argonne National Laboratory Contract No: DE-AC0206CH11357.

REFERENCES

1. Xu, T. T., Zheng, J. G., Nicholls, A. W., Stankovich, S., Piner, R. D., and Ruoff, R. S., "Single-Crystal Calcium Hexaboride Nanowires: Synthesis and Characterization," Nano Letters Vol. 4, No.10 pp. 2051-2055, August 17, (2004).

2. Shao, Y., Yin, G., Wang, Z., Gao, Y., "Proton Exchange Membrane Fuel Cell from low temperature to high temperature," Journal of Power Sources, 167 (2007), pp. 235-242.

3. Zheng, S., Min, G., Zou, Z., Yu, H., and Han, J., "Synthesis of calcium hexaboride powder via the reaction of calcium carbonate with boron carbide and carbon," J. Am. Ceram. Soc., **84** [11] 2725-27, (2001).

4. Glatzmaier, Gregory C., and Koc, R., "Method for silicon carbide production by reacting silica with hydrocarbon gas", June 28, 1994, U.S. #5324494.

5. Glatzmaier, Gregory C., and Koc, R., "Process for synthesizing titanium carbide, titanium nitride, and titanium carbonitride," May 23. 1995, U.S. #5417952.

6. Chung, F., Smith, D., "Industrial Applications of X-Ray Diffraction" Marcel Dekker, Inc. 2000. ISBN: 0-8247-1992-1.

7. Mwakikunga, B.W.; Sideras-Haddad, E.; Arendse, C.; Witcomb, M.J.; Forbes, A., "WO$_3$ Nano-Spheres into W$_{18}$O$_{49}$ One-Dimensional Nano-Structures Through Thermal Annealing", Journal of Nanoscience and Nanotechnology, Volume 9, Number 5, May 2009 , pp. 3286-3294(9).

SYNTHESIS OF SUBMICRON/NANO SIZED CaB$_6$ FROM CARBON COATED PRECURSORS

Naved Siddiqui[+], Rasit Koc[*]
Mechanical Engineering & Energy Processes
Southern Illinois University Carbondale, Carbondale, IL – 62901, USA
+Currently at Materials Engineering, Auburn University, Auburn, AL – 36849, USA
*Corresponding Author (E-mail: kocr@siu.edu, Tel: 618-453-7011)

Jennifer Mawdsley, David Carter
Argonne National Laboratories
Argonne, IL – 60439, USA

ABSTRACT

Calcium Hexaboride has been successfully synthesized at the laboratory scale using carbon coated precursors. Carbon coating is performed via pyrolysis of propylene at 550°C in a coating reactor on precursors. Various sets of precursors have been prepared where the starting materials include Calcium Carbonate, two different sources of Boron, and Carbon-black. An effort has been made to study the effects of various amounts of coatings on different combinations of starting materials to form the precursor material, and characterize the CaB$_6$ formed following the carbothermal reduction of these carbon coated powders under continuous argon flow at a maximum temperature of 1600°C with a holding time of 4 hours. XRD analysis shows that single phase CaB$_6$ can be successfully formed using carbon coated precursors while employing either boron source. TEM imaging and BET surface area measurements indicate that carbon coated precursors provides CaB$_6$ powders with fine particles and uniform morphology. The results and discussion provide emphasis on approaches for preparing carbon coated precursors depending on the boron source.

INTRODUCTION

Calcium hexaboride, a cubic metal boride belonging to the group 2A is a material attracting much attention due to its high hardness, high melting point, low density, chemical stability, and high electrical conductivity [1-3]. It has been reported as a material that can be exposed to high temperatures, while being able to provide surface protection in corrosive environments. Various studies have investigated the synthesis of CaB$_6$ powders using various methods as reported in [4-6], where researchers have employed the boron carbide method to produce CaB$_6$. However, the boron carbide method so far produced micron sized particles. Other studies have also used chlorine based compounds for CaB$_6$ synthesis, but these are thought to be rather dangerous experiments [7-8].

The lightweight, high density and highly conductive properties of calcium hexaboride make it a potential candidate for use in PEM Fuel Cells as a filler material for bipolar plates. Since bipolar plates account for 80% of the weight of a fuel cell stack, and 45% of stack cost [9, 10]; high quality CaB$_6$ produced at low cost can potentially fulfill these responsibilities.

The goal of this study is to illustrate the synthesis of CaB$_6$ using carbon coated precursors, formed via pyrolysis of propylene. This low cost method developed by Koc and Glatzmier, as briefly mentioned in [11], provides a pure form of carbon, which is amorphous, and provides excellent overall contact with the reacting powder. Various carbon coated precursors have been developed and described in the Results section. Two sources of boron have been employed in this study – Boron Source A and B, and it is very interesting to see how the dynamics for preparation of precursors and carbon coating change for the production of these precursors for synthesis of CaB$_6$. The details on the sources of Boron, and in depth analysis on effects will be published elsewhere, and cannot be mentioned here due to proprietary nature of the work under current contract and pending patents.

MATERIALS & METHODS

Coating Process

The preparation of precursors using the carbon coating process was carried out using a *Carbolite Rotary Rector* furnace. A predetermined amount of powders or mixture of powders were loaded into a specially designed stainless steel vessel, and affixed on the furnace. The vessel was flushed with argon and evacuated using a vacuum pump multiple times in order to create an inert atmosphere. Once this was done, a low concentration of argon (about 20 psi) was introduced into the vessel, and the temperature was ramped up to 550°C. Whenever the pressure in the vessel went above 35 psi, a small amount of argon was let out from the outlet. Once the temperature reached 550°C, propylene (C$_3$H$_6$) was inlet in the system, which was the carbon coating gas. At 550°C, propylene has been described to pyrolize or crack, and deposit a low density, amorphous and highly porous carbon on the surface of the powder loaded in the reactor [11]. Propylene was inlet up to a pressure of about 30 psi, and allowed to react with powders for 20 minutes, which was counted as 1 cycle of coating. After the 20 minute cycle, propylene was evacuated from the tube, and a fresh batch of propylene was inlet. After every third cycle, the reactor tube vessel was flushed with argon, before the propylene was inlet for the following cycle. After the last cycle, propylene is evacuated and the powders are allowed to cool down under argon. Therefore, on various powders, various numbers of coating cycles have been performed and discussed.

Synthesis of Coated Powders

Carbon coated precursors were drawn from the coating reactor, and if required, they were mixed with additional powders such as carbon black in a polystyrene vial, and milled in a Spex 8000 miller for 20 minutes while utilizing methyl methacrylate balls as the milling media.

The final mixed powders of a predetermined weight were placed in an open top graphite crucible, and placed in a high temperature box furnace (CM Inc. Rapid Temp 1704 Series). The mixed powders were then ramped to a desired temperature (1600°C, 1500°C, 1400°C) at a heating and cooling rate of 4°C/min under continuous argon flow. After a desired holding time (usually 4 hours), the powders were drawn and weight loss was determined. They were then collected and stored in a polystyrene vial for further characterization and analysis.

Characterization of Powders

Precursors or synthesized powders were characterized using X Ray Diffraction, performed in a *Rikagu Miniflex* using Cu Kα radiation at 2°/min from 20° to 70°. Using X-Ray diffraction (XRD) patterns, it was determined whether if the synthesized powders were single phase CaB$_6$ or not. In the event they were not, intermediate compounds could be readily seen. BET (Brunauer-Emmett-Teller) Surface Area of precursors, and synthesized powders were performed using a Micromeritics Gemini 2360 using N$_2$ adsorption at 77K. Samples were degassed at about 165°C overnight prior to the surface area analysis. The surface area measurements gave an excellent idea on the characteristics of the synthesized powders, as lower the surface area, the larger the CaB$_6$ particles were thought to be. The morphology of the powders, and average particle sizes were determined using Transmission electron microscopy (TEM) imaging, conducted a Hitachi H7650. EDS was also performed on some samples, which was done using a Hitachi S2460N.

RESULTS & DISCUSSION

The results presented in this section are divided into two sections, which are mainly based on the source of boron used for the preparation of precursor materials for the synthesis of calcium hexaboride. A brief explanation is given on various precursors that were developed for synthesis of CaB_6, and the most successful results have been illustrated.

Preparation of CaB_6 using the Boron Source A (B_A)

Calcium Hexaboride was first prepared using a stoichiometric equation, based on which the starting materials were Boron Source A, Calcium Carbonate, and Carbon. The calcium carbonate used was obtained from *Minerals Technologies* and was of the *Vicality Extra Light* variety. Carbon-black, if used was obtained from *Cabot Corporation*, and was of the *Monarch 880* type. However, the main source of carbon desired to be pyrolized carbon from propylene.

The precursor prepared for the synthesis of CaB_6 included Calcium Carbonate and Boron Source A (B_A). An experiment was done where the mixture of the two compounds was coated with propylene using the coating process over a number of cycles, but the resulting precursor had a substantial carbon deficiency. In order to accommodate for this, a subsequent precursor was prepared using a mixture in the required amounts of B_A, Calcium Carbonate and Carbon-black. This mixture was milled and coated with 20 cycles of carbon using propylene cracking at 550°C in the reactor tube.

The prepared precursor was synthesized at 1600°C, 1500°C and 1400°C under flowing argon for 4 hours, and XRD spectra shown in Figure 1 from this batch showed successful synthesis of CaB_6 at 1600°C. There were some intermediates found at 1500°C, while synthesis at 1400°C clearly showed an incomplete reaction with many intermediates present. The weight losses, BET Surface Area, and crystallite site calculated using Scherrer's equation are shown in Table 1. The corresponding XRD spectra of the three samples are shown in Figure 1.

Table 1: Results from CaB_6 synthesized using ($CaCO_3$, B_A, and C-black) coated for 20 cycles

Reaction Temp (°C)	Weight Loss (%)	BET Surface Area (m^2g^{-1})	Crystallite Site (nm)
1600	38.69	3.50	53
1500	35.03	3.96	54
1400	29.43	8.04	25

The XRD spectra in Figure 1 indicates that CaB_6 was successfully synthesized when treated at 1600°C as the spectrum matches very well with the one in literature and shows a single phase material. The same can be said about the synthesis at 1500°C as well. However, in the 1500°C sample, at lower 2-θ angles, there is a small presence of peaks that are not attributed to calcium hexaboride, and indicate the presence of tiny amounts of un-reacted boron. Another data for comparison is the weight loss, which happens to increase as the reacting temperature is increase as shown in the table. Even though the XRD spectra show a relatively complete reaction, the weight loss is below the 45% as targeted by stoichiometric data from the HSC software package. These values at 39% and 35% for 1600°C and 1500°C reactions respectively. In addition, the similarity in the specific surface area, which is about $3.5 - 4$ (m^2g^{-1}) for the two mentioned samples indicating that the synthesis of Calcium Hexaboride at these temperatures is relatively similar. As the reaction temperature is at 1400°C however, the specific surface area is much larger at 8.04 (m^2g^{-1}). This clearly shows that single phase formation of CaB_6 is not present at the reaction temperature of 1400°C, as it is at the higher temperatures. A holding time longer than 4 hours may be required for this reaction to completely take place. However, from all the results, the 1600°C sample from this precursor gave the most intense XRD spectrum for single phase

CaB$_6$. TEM imaging of the 1600°C sample prepared from this precursor is shown in Figure 2a and 2b, showing relatively uniform particle morphology. The particles from the images appear to be about 500 nm in diameter. However, the crystallite sites calculated using Scherrer's formula from the 110 peak give values that are an order of magnitude lower, showing that they are roughly 50 nm in size describing very good quality CaB$_6$. However, difference between the calculated particle size and the image gives a strong indication that the material formed is highly agglomerated, even after careful sample preparation prior to TEM imaging. This suggests that calcium hexaboride is successfully formed at 1600°C, but a longer reaction time may be required to break down the material further for finer morphology, and uniform particle distribution with fine particles, specially at the lower reaction temperatures.

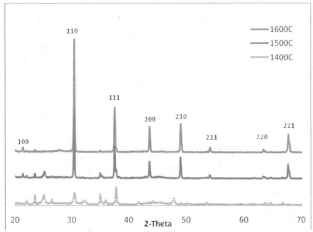

Figure 1 - XRD Spectrum of CaB$_6$ prepared from (CaCO$_3$ + B$_A$ + C-black) coated for 20 cycles

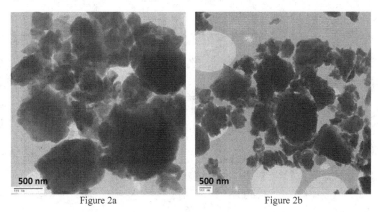

Figure 2a   Figure 2b

Figure 2: TEM of CaB$_6$ from (CaCO$_3$ + B$_A$ + C-black) coated for 20 cycles at 1600°C

Preparation of CaB$_6$ using Boron Source B (B$_B$)

In this section of research, studies were conducted to prepare CaB$_6$ by employing Calcium Carbonate, Boron Source B (B$_B$) and Carbon-black. The source of Boron used in this case is a cheaper source, and also allows the formation of CaB$_6$ at lower temperatures; hence the drive was to study this material in more detail. Another drive was to study, emphasize and quantify the effect of carbon coating on the precursor. After many experiments, it was determined that B$_B$ is not a material fit for coating, which was attributed to its lower melting temperature. Therefore, the effect of coating was studied on calcium carbonate, which was then mixed with the boron source and carbon black in required amounts based on the stoichiometric reaction.

Three batches of precursors were prepared based on the varying amounts of number of cycles of carbon coating on calcium carbonate. One batch had non-coated-CaCO$_3$, the second batch had 10 cycle-carbon-coated CaCO$_3$, and the third batch had 30-cycle-carbon-coated CaCO$_3$. Each of these were mixed with B$_B$ and carbon black. Carbon content analysis on each of the calcium carbonate varieties was performed at *Leco Corporation*. The results for this are provided in Table 2, which show an expected decrease in surface area upon the increase in carbon content. The carbon content in the non-coated calcium carbonate is from the carbon present in the material itself; and as the number of carbon coating cycles are increased, the carbon content % is increased. However, even after 30 cycles of coating, the amount of carbon deposited was raised only up to about 20% from the originally present 11%, and was not substantial based on the requirement posed by the stoichiometric reaction. Therefore, carbon black had to be added to the precursor mixture along with Boron source B (B$_B$).

Table 2: Effect of Carbon coating on Calcium Carbonate

	BET Surface Area (m^{2}g^{-1})	Carbon content (%)
Non-coated CaCO$_3$	8.96	11.4
10x-coated CaCO$_3$	5.80	14.6
30x-coated CaCO$_3$	4.95	19.8

X-ray diffraction spectra of Calcium Hexaboride prepared from the three batches of precursors prepared at 1600°C are provided in Figure 3, while the quantitative data is provided in Table 3. The discussion followed also includes description of TEM images of these samples, which are provided in Figures 4, 5 and 6 show CaB$_6$ prepared from non-coated CaCO$_3$ based precursor, 10-cycle-coated-CaCO$_3$ based precursor, and 30-cycle-coated-CaCO$_3$ based precursors respectively, all synthesized at 1600°C.

Table 3: Results from CaB$_6$ prepared using CaCO3 + BB + C-black synthesized at 1600°C

Precursor	Weight Loss (%)	BET Surface Area (m^{2}g^{-1})	Crystallite Site (nm)
Non-coated-CaCO$_3$ + B$_B$+C-black	84.16	0.98	55
10x-coated-CaCO$_3$+B$_B$+C-black	82.34	2.09	39
30x-coated-CaCO$_3$+B$_B$+C-black	83.32	4.05	39

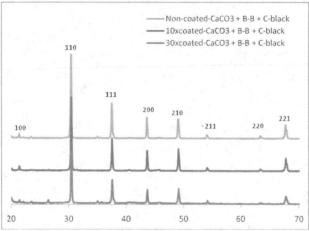

Figure 3: Effect of carbon coating on CaCO$_3$ for preparation of CaB$_6$ at 1600°C

The XRD spectra shown in Figure 3 compares Calcium Hexaboride prepared using a mixture of B$_B$, Carbon black, and three varieties of calcium carbonate. It can be seen that the most intense 110 peak of CaB$_6$ occurs using non-coated calcium carbonate, which indicates strong completion of the reaction, which is reaffirmed by the weight loss after the reaction, which was 84.16%, greater than the required 82.60%. However, here, a few tiny peaks which are not indexed are found at 2-theta angles of around 37°, which indicates the presence of boron, which did not react completely from the precursor material to form calcium hexaboride. In addition, TEM analysis of this sample, provided in Figures 4a and 4b also show that, on average most particles are about 1 micro meter in diameter. This however is not in agreement with the crystallite sites found using Scherrer's equation, which indicate nano-sized particles ranging around 55 nanometers, thus indicating the presence of strong agglomerates. Figure 4a indicates that particle shape is not well defined here, even though there were a few agglomerates of particles of oval or random shapes, which can be seen in Figure 4b. In addition, the specific surface area of the non-coated CaB$_6$ was found to be 0.98 m^2 g^{-1}. The extremely low surface area of less than 1 m^{2}g^{-1} also indicates a larger particle size relative to the other samples in this series.

As we compare the results with the 10 cycle coated calcium carbonate based precursor, we find that the 110 peak is not as intense as in the case of the non-coated CaCO$_3$ precursor, but the overall XRD spectra indicates a more single phased material, as there are no visible non-indexed peaks. All the peaks match very well with the spectrum found in various reported studies. The 1600°C reaction for 3 hours also shows a weight loss of 82.34%, which indicates near completion of the reaction, as it is very close to the required 82.60% from stoichiometry. Numerous TEM images were collected which show that most particles were below 500 nm in size, as it can be seen in Figure 5b. Figure 5a shows some larger particles, however the morphology of these larger particles are showing a very high agglomeration in the plate like structures. However, agglomeration again seems to be an issue as the crystallite site calculated is found to be 39 nm for the samples prepared using the precursor containing the carbon coated precursors. The specific surface area of this specimen was found to be 2.09 m^2 g^{-1}, slightly larger than the previous non-carbon-coated sample, and therefore this gives affirmation that the particle size is in fact smaller than before. Therefore, there is evidence of smaller particles and better morphology, but high agglomeration remains an issue.

Upon analysis of the sample containing the 30-cycle-coated-CaCO$_3$+B$_B$+C, we find that the XRD spectra does not show single phase Calcium Hexaboride, indicating that the formation of the material is not single phased. This sample contains the highest amount of carbon, as it contains the same amount of carbon black in addition to the 30 cycle carbon coating as compared to the other samples. The main reason that this sample did not produce single phase CaB$_6$ is most likely due to a deficiency of calcium in the CaCO$_3$ precursor, which may be hindering the reaction mechanism based on the required amounts from stoichiometry. The weight loss from the 1600°C reaction was 83.32%, which does indicate completion of the reaction, but the XRD spectra is not single-phased due to the presence of peaks not attributed to CaB$_6$. Also, the TEM imaging shows that there is a large particle size distribution in this sample, especially in Figure 6b. The morphology of the particles is also not very well defined, as we see more cloud like structures in Figure 6b; but we do see a number of small particles as well. The specific surface area of this sample was found to be 4.06 m^{2}g^{-1}, which shows a comparatively larger surface area. Therefore, with a higher amount of carbon coating, it seems that the particle sizes can be smaller, but the result is not entirely convincing due to the deficiency of CaCO$_3$. Otherwise, this batch is expected to provide the best quality CaB$_6$.

Figure 4a: TEM of non-coated CaCO$_3$ based CaB$_6$

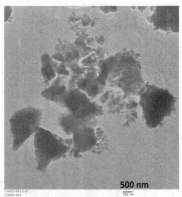

Figure 4b: TEM of non-coated CaCO$_3$ based CaB$_6$

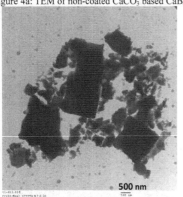

Figure 5a - TEM of 10xcoated-CaCO$_3$ based CaB$_6$

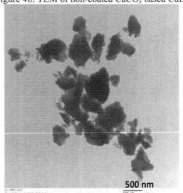

Figure 5b - TEM of 10xcoated-CaCO$_3$ based CaB$_6$

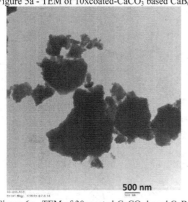

Figure 6a - TEM of 30xcoated-CaCO$_3$ based CaB$_6$

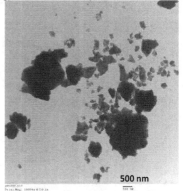

Figure 6b - TEM of 30xcoated-CaCO$_3$ based CaB$_6$

Following these experiments, it was desired to react the above mentioned precursors at 1500°C under flowing argon. The results from this batch are provided in Table 4, while the XRD spectra are provided in Figure 7.

Table 4: Results from CaB$_6$ prepared using CaCO3 + BB + C-black synthesized at 1500°C

Precursor	Weight Loss (%)	BET Surface Area (m^{2}g^{-1})	Crystallite Site (nm)
Non-coated-CaCO$_3$ + B$_B$+C-black	85	1.04	45
10x-coated-CaCO$_3$+B$_B$+C-black	84	3.94	39
30x-coated-CaCO$_3$+B$_B$+C-black	78	3.63	34

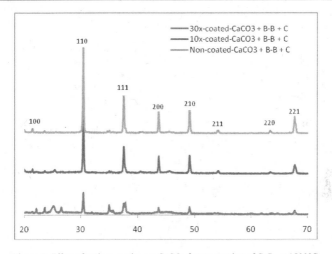

Figure 7: Effect of carbon coating on CaCO$_3$ for preparation of CaB$_6$ at 1500°C

The significance of the above mentioned experiments is the successful synthesis of calcium hexaboride, at 1500°C, as it can be seen that the non-coated-CaCO$_3$ + B$_B$ + C precursor yielded a single phased XRD spectra when treated at 1500°C for 4 hours under flowing argon in the box furnace. The sample coated with 10 cycle- CaCO$_3$ + B$_B$ + C also showed formation that can be considered single phase, due to the absence of any major peaks that are not attributed to Calcium Hexaboride. However, in the case of the 30-cycle-coated-CaCO$_3$ + B$_B$ + C, treated at 1500°C, it is found that the formation is definitely not single phased, and shows CaB$_6$ peaks, but not very intense at all. The specific surface area of these samples are 1.04 m^2 g^{-1}, 3.95 m^2 g^{-1}, and 3.63 m^2 g^{-1} for the three samples in order of coating respectively. From the surface areas as well, it is quite evident that the non-coated sample shows a stronger completion of the reaction, but the higher surface area in case of the coated sample, especially the 10-cycle coated-CaCO$_3$ sample, indicates better particle size. Therefore, the findings with the lower temperature reactions are very similar to the ones performed at 1600°C, indicating that the potential calcium carbonate deficiency is playing a major role in hindering the formation of single phase CaB$_6$. Therefore, the crystallite site calculation indicating a site of 34 nm for the 30-cycle-calcium-carbonate coated sample could be misleading, as this does not indicate that this crystallite site is of calcium hexaboride, as the XRD spectra clearly indicates otherwise. This particle size may be more attributed towards an oxide material.

As an overall note from this set of experiments, we can clearly see that the formation of single-phase CaB_6 is indeed possible at 1500°C, as it is seen in the lower concentration of carbon content samples with the non-coated-$CaCO_3$ based sample, and the 10-cycle-coated $CaCO_3$ based sample. Further experiments in order to prove the synthesis of calcium hexaboride using the 30-cycle-coated-$CaCO_3$ + Boron Source B + Carbon-black are needed. In order to achieve this, calcium carbonate deficiency should be carefully taken into account.

Seeding of 10-cycle-coated-$CaCO_3$ + Boron Source B + Carbon-black

Following the set of experiments described in the last section, an effort was made to improve the morphology of the particles with the goal of obtaining a higher aspect ratio in the particle shape. This was done by seeding the 10-cycle-coated-$CaCO_3$ + B_B + C precursors by adding an additional 1% Seeding Element A (S_A) in one case and 1% Seeding Element B (S_B) in another case, both being transition metals. S_A and S_B were added to the mentioned precursor in an argon-flushed glove-box, after which they were mixed for 20 minutes in the *Spex 8000* mixer. These samples were reacted at 1600°C for 3 hours. The results data is provided in Table 5. XRD spectra of the seeded samples are provided in Figure 8 and TEM imaging is illustrated in Figures 9 and 10.

Table 5: Results from CaB_6 prepared using 10x-coated-$CaCO_3$ + B_B + C-black + Seeding Element synthesized at 1600°C

Precursor	Weight Loss (%)	BET Surface Area (m^2g^{-1})	Crystallite Site (nm)
10x-coated-$CaCO_3$ + B_B+ C-black + S_A	84.42	2.59	46
10x-coated-$CaCO_3$ + B_B+ C-black + S_B	90.49	3.55	46

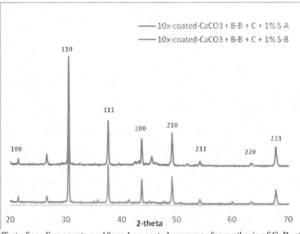

Figure 8: Effect of seeding agents on 10xcarbon coated precursor for synthesis of CaB_6 at 1600°C for 3 hours

The XRD spectra of the two seeded samples show that the final product is primarily CaB_6, along with a few extra peaks. The closest matches that were found for the other peaks, present around 2-theta angles of about 28° were carbides of the transition metals. However, the intensities of these peaks were too low to confirm this. This is because the addition of 1% of these powders is quite small to show any significant formation of other compounds, and therefore, the goal of seeding them using these particles is kept in check. EDS spectra of these samples were collected, which indicated the presence of the respective transition metals. Boron happens to be a much lighter element, and seems to be rather unresponsive to EDS. Therefore, the peaks primarily show two peaks of Calcium, which indicate kα_1 and kβ_1. But, from XRD analysis, we know for a fact that the compound is primarily CaB$_6$.

However, after a closer look at the XRD spectrum, the sample prepared with 1% S$_A$ shows a the presence of a few more peaks between the 200 and 210 calcium hexaboride peaks, compared to the sample with 1% S$_B$. TEM imaging of these samples provided in Figures 9 and 10 show this is more detail. It can be clearly seen from the differences in these two figures that S$_A$ has a much stronger effect on the morphology of calcium hexaboride. Overall, in general, the particles looked like the ones provided in Figure 9a, which showed some small sub-micron cloud like structures, along with the presence of elongated structures, which looked more like rods or wires. The morphology in this case is more defined. Figure 9b shows a magnified view of the rod-like structure. The weight loss of this sample was larger than the non-seeded sample, at 84.42%. The specific surface area also increased on comparison at 2.60 m^2 g^{-1}, indicating that the overall size of particles was smaller than the non-seeded sample. However, the crystallite site calculation shows that the site was about 46 nm, comparable to the non-seeded counterpart, which was 39 nm.

The TEM imaging of the 1% S$_B$ sample does not show much change in the morphology of the samples. However, there is a general feel that the particles from this reaction are more consistent, as it can be seen in Figures 9a and 9b. The particle sizes are mostly sub-micron from this sample, but there were a few that were largely agglomerated, as evident from the crystallite site calculation, and have not been included here. The weight loss after the reaction of this sample was found to be quite large though, with only a 9.51% final yield of CaB$_6$. The specific surface of this sample was larger than the 1% S$_A$ and the non-seeded 10-cycle-coated-CaCO$_3$+B$_B$+C at 3.52 m^2 g^{-1}.

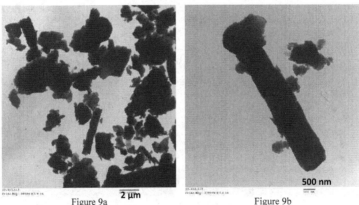

Figure 9a   Figure 9b
Figure 9: TEM Imaging of 10-cycle-coated-CaCO$_3$ + B$_B$ + 1% S$_A$

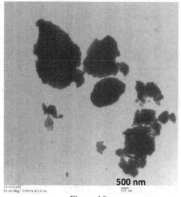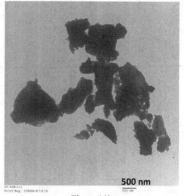

Figure 10a   Figure 10b

Figure 10: TEM Imaging of 10-cycle-coated-CaCO$_3$ + B$_B$ + 1% S$_B$

CONCLUSIONS

The differences between the two sources of boron and the various methods employed for the preparation of precursors yield some interesting inferences. Boron source A seems to be a much tougher material to break down for the synthesis of CaB$_6$, and is a much more expensive product compared to Boron Source B. However, the yield of CaB$_6$ form this particular method is much greater than compared to the Boron Source B reaction, which happens to be a big advantage. Another advantage that Boron Source A provides is that it is a material that can be coated unlike Source B, which melts while coating with propylene. It may be possible to eliminate the use of carbon black completely during the preparation of precursor from B$_A$ due to the comparatively lower amount of carbon required as an input. This however indicates that the quality of B$_A$ being used, which is commercially acquired becomes very important. For high quality CaB$_6$, it would be very important to use the best possible B$_A$, which may turn out to be quite an expensive proposition. Nevertheless, a completely carbon black free precursor could be prepared using carbon-coated-calcium-carbonate, mixed with B$_A$. This mixture could then be further coated with carbon again using propylene cracking. This is a proposal for a future precursor material.

In the case of Boron Source B, it would be very difficult to obtain a completely carbon black free precursor due to the high carbon requirement, in addition to the fact that the boron source cannot be coated, and only that calcium carbonate is coated and mixed with the other requirements. However, the driving force behind using B$_B$ is its low cost, and also as shown from the experiments and TEM imaging the high quality yield of CaB$_6$ from these precursors, and a lower possible reaction temperature for completion of reaction for producing calcium hexaboride. An overall carbon coated precursor could perhaps be obtained by using a mixture of carbon coated calcium carbonate mixed with B$_B$; and carbon black amounting to half of the total required carbon. This mixture would be milled and then further coated substantially to meet the carbon requirement. But, B$_B$ is an attractive material to study due to the lower temperature required to synthesize CaB$_6$, which would be a big cost factor.

It is also noted from the shown experiments that calcium carbonate can be easily coated to meet some amount of the carbon requirement in each case. However, in doing so, it must be made sure that there is no calcium deficiency in the calcium carbonate upon coating, and the starting materials meet stoichiometric requirements as it was found in the case of the Boron Source B reactions with a higher number of carbon coatings. It was however shown that the coated powders are providing better morphology, due to the better contact of carbon with calcium carbonate, and also the more pure form of carbon as compared to carbon black, it provides a very attractive proposition for formation of high quality CaB$_6$.

Seeding was another issue addressed during research, and it was seen that adding 1% of S$_A$ to the overall mixture provided better shaped particles than the 1% S$_B$ mixture. However, the results were mixed and further research needs to be conducted to determine proper holding times, and flow rates for production of seeded CaB$_6$. S$_A$ and S$_B$ could also be used mixed in the powders prior to the coating process and tested. Overall, the experiments provide an excellent preliminary background for further experiments and synthesis of high quality CaB$_6$.

ACKNOWLEDGEMENTS

This research has been performed at Southern Illinois University at Carbondale, sponsored by the U.S. Department of Energy through Argonne National Laboratory under contract number: DE-AC0206CH11357. The researchers are thankful for the support.

REFERENCES

[1] T. Xu, J. Zheng, A. Nicholls, S. Stankovich, R. Piner, R. Ruoff, Single-Crystal Calcium Hexaboride Nanowires: Synthesis and Characterization, *Nano Letters*, **4**, 2051-2055 (2004)

[2] J. Matsushita, K. Mori, Y. Nishi, Y. Sawada, Oxidation of Calcium Boride at High Temperature, *Journal of Materials Synthesis and Processing*, **6**, 407-410, (1998)

[3] S. Zheng, G. Min, Z. Zou, S. Tatsuyama, High Temperature Oxidation of calcium hexaboride powders, *Materials Letters*, **58**, 2586-2589, (2004)

[4] S. Zheng, G. Min, Z. Zou, H. Yu, J. Han, Synthesis of Calcium Hexaboride Powder via the Reaction of Calcium Carbonate with Boron Carbide and Carbon, *Journal of American Ceramic Society*, **84**, pp. 2725-2727 (2004)

[5] Z. Lin, M. Guanghui, Y. Huashun, Reaction mechanism and size control of CaB$_6$ micron powder synthesized by the boroncarbide method, *Ceramics International*, **35**, 3533-3536, (2009)

[6] L. Zhang, G. Min, H. Yu, H. Chen, G. Feng, The Size and Morphology of Fine CaB$_6$ Powder synthesized by Nanometer CaCO$_3$ as Reactant, *Key Engineering Materials Vols.*, **326-328**, 369-372, (2006)

[7] L. Shi, Y. Gu, L. Chen, Z. Yang, J. Ma, Y Qian, Low Temperature Synthesis and Characterization of Cubic CaB$_6$ Ultrafine Powders, *Chemistry letters*, **32**, 958-959, (2003)

[8] J. Xu, Y. Zhao, C. Zou, Q. Ding, Self-catalyst growth of single-crystalline CaB$_6$ nanostructures, *Journal of Solid State Chemistry*, **180**, 2577-2580, (2007)

[9] A. Hermann, T. Chaudhari, P. Spagnol, Bipolar Plates for PEM Fuel Cells: A review, *International Journal of Hydrogen Energy*, **30**, 1297-1302, (2005)

[10] D. Hodgson, B. May, P. Adcock, D. Davis, New lightweight bipolar plate system for polymer electrolyte membrane fuel calls, *Journal of Power Sources*, **96**, 233-235 (2001)

[11] R. Koc, J. Folmer, Carbothermal synthesis of titanium carbide using ultrafine titania powders, *Journal of Materials Science*, **32**, 3101-3111 (1997)

AN EASY TWO-STEP MICROWAVE ASSISTED SYNTHESIS OF SNO₂/CNT HYBRIDS

Sarah C Motshekga[1, 2], Sreejarani K Pillai[1]*, Suprakas Sinha Ray[1,] Kalala Jalama[2], Rui.W.M Krause[2]

[1]DST/CSIR Nanotechnology Innovation Centre, National Centre for Nano-Structured Materials, Council for Scientific and Industrial Research, Pretoria 0001, Republic of South Africa
[2] University of Johannesburg, Department of Chemical Engineering, Doornfontein, 2028, South Africa.

ABSTRACT

Tin oxide (SnO₂) - decorated carbon nanotube (CNT) heterostructures were synthesized by microwave assisted wet impregnation method. CNTs of three different aspect ratios were compared. The hybrid samples were characterized by powder X-ray diffraction, Raman spectroscopy, high resolution transmission electron microscopy, BET surface area analysis and DC conductivity measurement. The results showed that the microwave assisted synthesis is a very efficient method in producing CNTs that are heavily decorated by SnO₂ nanoparticles in a very short time (total reaction time of 10 min.), irrespective of their length and diameter. The hybrids showed 100 times increase in electrical conductivity when compared to the unmodified CNTs.

INTRODUCTION

Hybrid structures of nanoparticles (NPs) distributed on CNT surface could potentially display not only the unique properties of nanocrystals and those of CNTs, but also additional novel properties. Attaching NPs on CNTs surfaces could promote direct and effective charge transfer and hence increase device/process efficiency.[1] SnO₂ NPs decorated CNTs are reported to be useful functional composites in many applications including gas sensors[2], fuel cells[3], batteries[4], and supercapacitors.[5] This is based on the fact that the work function of CNTs is approximately equal to that of SnO₂ allowing electrons to travel through the SnO₂ grains to CNTs and then be conducted in the CNTs with low resistance.[6] Over the last few years various techniques have been used to prepare SnO₂/CNT hybrid structures such as wet-chemical[7-10], sol-gel[11], gas-phase[12, 13], supercritical fluid[14] methods, etc. Bai et al.[6] reported a microwave–polyol irradiation method for fuctionalizing CNTs with SnO₂ where diethylene glycol was used as the solvent. Prior to the microwave irradiation, the CNTs were oxidized in concentrated HNO₃ at 140°C for 6 h. Microwave assisted synthesis has recently shown remarkable advantages over the conventional synthesis routes such as rapid volumetric heating, high reaction rate, reduced particle size, homogenous and narrow size distribution of particles.[6, 15] In this study, we report a fast and efficient microwave-assisted two-step synthesis for coating CNTs with SnO₂ NPs.

EXPERIMENTAL

Three batches of Multi-Walled Carbon Nanotubes (MWCNTs) produced by chemical vapor deposition (CVD) method were provided by Sigma-Aldrich and correspondingly abbreviated as CNT10 (OD =10 30 nm, ID=3 10 nm, L = 10 m, 90% purity), CNT200 (OD = 20 30 nm, ID = 5 10 nm, L = 200 m, 95% purity), and CNT500 (OD = 40 60 nm, ID = 5 10 nm, L = 500 m, 95% purity). A portion of CNTs (500 mg) were heated under reflux with 5M HNO₃ (10 ml/ 10 mg) at 120°C for 5 min in a microwave reactor (Anton Paar microwave reaction system-Multiwave 3000) at a power setting of 500 W. The solution was filtered and washed with distilled water until the pH of the solution was neutral. The CNTs were then dried in air at 110°C for 12 h. In the second step, 10 g of SnCl₂ was dissolved in 400 ml of distilled water and 5 ml of concentrated HCl (37 wt%) was added. The acid treated CNTs were then added to the above solution. The solution was then heated under reflux in the microwave for 5 min at 500 W and 60°C. The precipitate was then separated from the mother liquor by centrifugation, washed with distilled H_2O several times and dried in air at 110°C for 12 h. The final product was calcined in air at 500 C for 2 h. The SnO_2/CNT composites obtained after microwave treatment with three different types of CNTs were correspondingly abbreviated as CNT10NC, CNT200NC and CNT500NC.

RESULTS AND DISCUSSION

Figure 1 presents a series of TEM images of composite samples at two different magnifications. All the three different batches of CNTs were observed to be heavily coated with SnO_2 NPs. These NPs were uniformly deposited even in the interior surface of the tubes and were spherical with 3-5 nm diameters.

The G/D ('G' for graphitic band and 'D' for defect band) ratios calculated from the Raman spectra for the CNTs and corresponding SnO_2-containing samples are presented in Table 1. The G/D ratio is greater for composites CNT200NC and CNT500NC compared to pure CNT200 and CNT 500, respectively. This could be due to an enhancement of atomic ordering of crystallinity of the CNTs after they were decorated with SnO_2 NPs [13]. The opposite was observed for CNT10 and its corresponding SnO_2-containing sample CNT10NC and could be due to the interaction between the SnO_2 and the surface group of CNTs rather than the increase in amorphous carbon.[16]

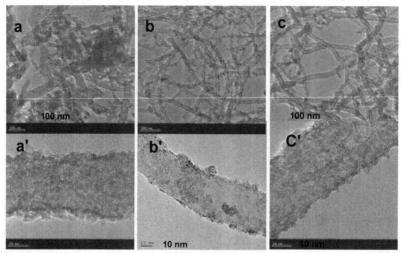

Figure 1. TEM images of various SnO_2/CNT composite samples at two different magnifications: (a & a') CNT10NC; (b & b') CNT200NC; (c & c') CNT500NC.

Table 1. Characteristic peak intensities and *G/D* ratios from the Raman spectra of CNTs and corresponding SnO$_2$-containing composite samples.

| Sample | Raman Intensities /a.u.[a] | | *G/D* ratio |
	G-band	*D*-band	
CNT10	16.51	16.68	0.99
CNT10NC	13.86	17.26	0.80
CNT200	14.77	16.52	0.89
CNT200NC	12.61	13.40	0.94
CNT500	15.02	16.67	0.90
CNT500NC	16.35	12.28	1.33

[a] '*G*' for graphitic band and '*D*' for defect band

To further confirm the presence of SnO$_2$ particles in the composite samples and check the crystalline phases, the prepared samples were analyzed by XRD and the diffractograms are presented in figure 2. All the samples show peaks of C (002) and C (100) phases with d$_{002}$ values of 0.34 nm characteristic of CNTs. The SnO$_2$/CNT samples show intense peaks corresponding to tetragonal phase of SnO$_2$. The C (002) peaks appear to be shifted in the composite sample indicating the interaction of SnO$_2$ nanoparticles with CNTs. XRD results are in line with microscopy observations.

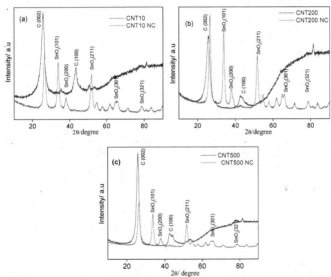

Figure 2. XRD patterns of pure CNTs and composite samples: (a) CNT10 & CNT10NC, (b) CNT200 & CNT200NC, and (c) CNT500 & CNT500NC.

The measured BET surface areas of different CNT samples are given in table 2. Surface areas of the SnO$_2$/CNT samples are observed to be lower than the corresponding pure CNTs in all the three batches. This indicates the coating of CNT surface with SnO$_2$ NPs which also contributes to the increase in pore size. The DC conductivity values of various samples obtained from four point probe measurements are also presented in table 2. The conductivity values increase appreciably for the composite samples. For example, the DC conductivity increases from 3.53×10^{-2} S/cm for the pure CNT200 to 3.23 S/ cm (average of five independent measurements with a maximum error of 7%) for the composite and this represents a 100 times increase in the electrical conductivity. These results indicate that there is effective contact between CNTs and the SnO₂ NPs forming a conducting network which helped in lowering the resistance to the conduction of electrons. Although high concentration of SnO₂ nanoparticles can not contribute to the electrical conduction[17-19], these results indicate that in the present case, the SnO₂ nanoparticle concentration is enough to form a percolating net work, which attributes to presence of high concentration of charge carriers[20] leading to increase in overall conductivity of the system.

On the other hand, the similar composites prepared by conventional wet impregnation method (results not shown here) showed only 10 fold increase in DC conductivity when compared to pure CNTs. This indicates the effectiveness of microwave assisted route for the preparation of such nanocomposites.

Table 2. Data obtained from BET and electrical conductivity measurements of various CNTs and SnO₂/CNT composite samples

Samples	Surface area /$m^2.g^{-1}$	Micropore volume /$cm^3.g^{-1}$	Pore size /nm	Electrical Conductivity/ S/cm
CNT10	129.0	0.4	11.7	2.16×10^{-2}
CNT10NC	103.2	0.4	15.4	8.28
CNT200	129.9	0.4	15.5	3.53×10^{-2}
CNT200NC	119.0	0.4	12.6	3.23
CNT500	129.7	0.4	11.5	2.02×10^{-2}
CNT500NC	107.6	0.5	13.5	4.01

CONCLUSIONS

We have described a simple, fast and efficient method to synthesize SnO₂ NPs decorated CNTs. The method serves equally good for the surface functionalization for CNTs with different aspect ratios. Nanoparticles of 3-5 nm were uniformly deposited on the CNT surface in a total reaction time of 10 min. The composites thus prepared exhibited 100 times increased electrical conductivity when compared to the pure CNTs. Our ongoing work will focus on the gas-sensing ability and catalytic efficiency of composite materials.

ACKNOWLEDGEMENTS

SCM, SKP and SSR thank the DST and the CSIR, South Africa for financial support.

REFERENCES

[1]G. Lu, M. Liu, K. Yu and J. Chen, Absorption Properties of Hybrid SnO₂ Nanocrystal-Carbon Nanotube Structures, *J. Electron Mater*., **37**,1686-90 (2008).
[2]Y. X. Liang, Y. J. Chen and T. H. Wang, Low-Resistance Gas Sensors Fabricated From Multiwalled Carbon Nanotubes Coated With a Thin Tin Oxide Layer, *Appl Phys Lett*., **85**, 666–8 (2004).
[3]K. Ke and K. Waki, Fabrication and Characterization of Multiwalled Carbon Nanotubes-Supported Pt/SnOₓ Nanocomposites as Catalysts for Electro-oxidation of Methanol, *J. Electrochem Soc*., **154**, A207-12 (2007).

[4]J. N. Xie and V. K. Varadan, Synthesis and Characterization of High Surface Area Tin Oxide/Functionalized Carbon Nanotubes Composite as Anode Materials, *Mater. Chem. Phys.*, **91**, 274–80 (2005).

[5] A. L. M. Reddy and S. Ramaprabhu, Nanocrystalline Metal Oxides Dispersed Multiwalled Carbon Nanotubes as Supercapacitor Electrodes, *J. Phys. Chem. C.*, **111**, 16138–46 (2007).

[6]J. Bai, Z. Xu and J. Zheng, Microwave-polyol Process for Functionalizing Carbon Nanotubes with SnO₂ and CeO₂ Coating, *Chem Lett.*, **35**, 96-7 (2006).

[7]J. G Zhou, H. T. Fang, J. M. Maley, J. Y. P. Ko, M. Murphy, Y. Chu, et al., An X-ray Absorption, Photoemission, and Raman Study of the Interaction between SnO₂ Nanoparticle and Carbon Nanotube, *J Phys Chem C.*, **113**, 6114-7(2009)

[8]W. Q. Han and A. Zettl, Coating Single-Walled Carbon Nanotubes with Tin Oxide, *Nano Lett.*, **3**, 681–3(2003).

[9] L. P. Zhao and L. Gao, Coating of Multi-walled Carbon Nanotubes with Thick Layers of Tin(IV) Oxide, *Carbon*, **42**, 1858–61(2004).

[10] L. P. Zhao and L. Gao, Filling of Multi-walled Carbon Nanotubes with Tin(IV) Oxide, *Carbon*, **42**, 3269–72 (2004).

[11]J. Gong, J. Sun and Q. Chen, Micromachined Sol-gel Carbon Nanotube/SnO₂ Nanocomposite Hydrogen Sensor, *Sens. Actuators B*, **130**, 829-35 (2008).

[12]L. Ganhua, E. O. Leonidas and C. Junhong, Room Temperature gas Sensing Based on Electron Transfer between Discrete Tin Oxide nanocrystals and Multiwalled Carbon Nanotubes, *Adv. Mater.*, **21**, 2487–91 (2009).

[13] F. Yanbao, M. Ruobiao, S. Ye, C. Zhuo and M. Xiaohua, Preparation and Characterization of SnO₂/Carbon Nanotube Composite for Lithium Ion Battery Applications, *Mater. Lett.*, **63**, 1946–8 (2009).

[14]G. An, N. Na, X Zhang, Z. Miao, S. Miao, K. Ding and Z. Liu SnO₂/Carbon Nanotube Nanocomposites Synthesized in Supercritical Fluids: Highly Efficient Materials for Use as a Chemical Sensor and as the Anode of a Lithium-ion Battery, *Nanotechnology*, **18**, 435707-13 (2007).

[15]C. Hao, Y. Du and L. J. Li, Microwave-Assisted Heating Method for the Decoration of Carbon Nanotubes with Zinc Sulfide Nanoparticles, *Dispersion Sci. Technol.*, **30**, 691–3 (2009).

[16] M. S. Park, S. A. Needham, G. X. Wang, Y. M. Kang, J. S. Park, S. X Dou, et al. Nanostructured SnSb/Carbon Nanotube Composites Synthesized by Reductive Precipitation for Lithium-Ion Batteries, *Chem. Mater.*, **19**, 2406–10 (2007).

[17] W. A. Gazotti, G. Casalbore-Miceli, A. Geri, A. Berlin and M. A. De Paoli, An All-Plastic and Flexible Electrochromic Device Based on Elastomeric Blends, *Adv.Mater.*, **10**, 1522–1525 (1998).

[18] M. J. Alam and D.C. Cameron, Investigation of Annealing Effects on Sol-gel Deposited Indium Tin Oxide Thin Films in Different Atmospheres, *Thin Solid Films*, **76**, 420–421 (2002).

[19] J. H. Hwang, P. P. Edwards, H. K. Kammler and T. O. Mason. Point Defects and Electrical Properties of Sn-Doped In-Based Transparent Conducting Oxides, *Solid State Ionics*, **129**, 135–144 (2000)

[20] Z. Qi, Z. Meifang, Z. Qinghong, L. Yaogang and W. Hongzhi, Fabrication and Characterization of Indium Tin Oxide–Carbon Nanotube Nanocomposites, *J. Phys. Chem. C*, **113**,15538-1143 (2009).

* Corresponding author. Tel.: +27128412646; Fax: +27 12 841 2229.
E-mail address: skpillai@csir.co.za (S. K .Pillai)

GRAIN SIZE REDUCTION AND SURFACE MODIFICATION EFFECT IN POLYCRYSTALLINE Y$_2$O$_3$ SUBJECT TO HIGH PRESSURE PROCESSING

Jafar F. Al-Sharab[1], Bernard H. Kear, S. Deutsch, and Stephen D. Tse
Center for Nanomaterials Research, Rutgers University, 607 Taylor Rd.
Piscataway, NJ 08854

ABSTRACT

Over the past decade, there has been growing interest in the fabrication of nanostructured oxide ceramics for diverse structural and functional applications. Typically, a nanostructured oxide is fabricated by pressure- or field-assisted sintering of a nanopowder compact. Even so, it is difficult to achieve complete densification without causing significant grain growth. In this paper, we describe a reversible-phase-transformation mechanism to convert full dense polycrystalline Y$_2$O$_3$ directly into the nanocrystalline state under high pressure and high temperature processing conditions (8GPa/1000 °C). In addition, we report a surface modification effect and superplastic compressive phenomena observed in polycrystalline Y$_2$O$_3$ processed at high temperature/pressure conditions.

INTRODUCTION

A major challenge encountered in consolidating an oxide ceramic nanopowder is to realize a nano-scale grain size in the final sintered product. In the early stages of sintering, rapid densification occurs driven by the high surface area of the nanoparticle compact. However, the grain size remains *small*, since a uniform distribution of nanopores in the incompletely densified ceramic inhibits grain growth. In the final stages of sintering (>90% theoretical density), when the nanopores begin to disappear, unrestricted grain growth occurs, typically leading to a *micro-grained* sintered product.

Two methods are available to overcome this limitation: one for single-component oxides (*nanocrystalline oxides*) [1-5] and the other for multi-component oxides (*nanocomposite oxides*) [6-10]. For a single-component oxide, the key to success is to start with a metastable nanopowder and to control its decomposition kinetics during high pressure sintering (HPS). Typically, the pressure required to obtain a fully dense nanocrystalline oxide is 3-8 GPa, which places a practical limit on the size of parts that can be fabricated using existing technology. For a multi-component oxide, the pressure requirement for powder consolidation is relaxed to 0.1-1.0 GPa, so that it is easier to fabricate larger parts. The reduced pressure requirement is due to the formation of a thermally-stable nanocomposite structure, in which one nanophase strongly impedes the growth of an adjacent nanophase(s), particularly when their volume fractions are comparable. In this paper, we are concerned primarily with single-component oxide ceramics, with the emphasis on polycrystalline Y$_2$O$_3$.

As noted above, the key to success in producing a fully dense nanocrystalline oxide ceramic is to control the decomposition kinetics of a metastable nanopowder during HPS. The effect has been observed for several oxide systems, but in a particularly striking form for a metastable anatase-TiO$_2$ nanopowder. At a pressure >1.5 GPa, the metstable anatase-TiO$_2$ phase transforms into the stable rutile-TiO$_2$ phase accompanied by rapid densification, apparently facilitated by the metastable-to-stable phase transformation during sintering. The resulting grain size is comparable to the starting nanoparticle size. This is in marked contrast to what happens during conventional hot pressing, where rapid grain growth occurs, leading to a micro-grained oxide.

The present paper describes a pressure-induced reversible phase transformation (RPT) process to transform bulk polycrystalline Y$_2$O$_3$ directly into the nanocrystalline state. As will be shown, processing involves a forward transformation from cubic-to-monoclinic under a high pressure and a backward transformation from monoclinic-to-cubic under a lower pressure, both steps performed at the same temperature and holding time. In addition, a surface modification effect, which includes the formation of surface columnar

---

[1] *Contact: jafarhan@rci.rutgers.edu*

grains, apparently suffered from superplastic compression flow due to high pressure/temperature processing will also be discussed. An on-going investigation is examining mechanisms and kinetics involved in RPT processing, with a view to establishing the optimal parameters (pressure-temperature-time) to achieve the finest possible nano-grain size in Y_2O_3, as well as selected Y_2O_3-base composites.

RESULTS AND DISCUSSION

Disc-shaped samples (4 dia. x 4 mm) of polycrystalline Y_2O_3, prepared by hot isostatic pressing (HIP) of powder compacts, were obtained from Raytheon IDS. The samples were subjected to various high pressure-high temperature (HPHT) treatments, using a high pressure unit of novel design [11], **Figure 1**. Pre-stressed 4340 steel rings encasing WC/6%Co anvils enable pressures up to 8 GPa, without causing anvil cracking. Resistive heating of a graphite crucible, which contains the sample, enables rapid heating up to the desired temperature. Pressure is calibrated via known data for pressure-induced phase transitions in Ce, Bi and PbSe. Temperature is calibrated via known values of melting points of Sn, Al and Cu under high pressure. In a typical HPHT experiment, the sample is introduced into the graphite crucible, subjected to high pressure, heated to high temperature, held for a specific time, and cooled under pressure. Heating and cooling rates are set at about 65°C/min.

SEM observation of a fractured sample showed that the grain size of the starting material was about 300 μm. XRD analysis showed that the material had the equilibrium cubic-Y_2O_3 structure. After various HPHT treatments, a *cubic-to-monoclinic* phase transformation invariably occurred. Upon XRD examination, it was apparent that this transformation was accompanied by a significant reduction in grain size. An example is shown in **Figure 2**, where a *forward* transformation from c-Y_2O_3 to m-Y_2O_3 at 1000°C/8GPa/15 minutes reduces the grain size from 300 μm to 50 nm, whereas a *reverse* transformation from m-Y_2O_3 back to c-Y_2O_3 at 1000°C/1GPa/15 minutes yields a grain size of ~100 nm. In other words, by a single reversible phase transformation, the grain size of the original micro-grained Y_2O_3 has been reduced by a factor of 3000.

As indicated in **Figure 2**, the surface-XRD pattern of the nanocrystalline c-Y_2O_3 has all the characteristics of a powder pattern, which indicates that the cubic nanoparticles must have experienced random nucleation during decomposition of the intermediate nano-monoclinic phase. In contrast, the XRD pattern of the initial micro-grained c-Y_2O_3 phase shows intensity distributions that are not consistent with a powder pattern, although all the expected peaks are visible. This is because only a few surface grains are in favorable orientations for diffraction.

Density measurements showed that the forward phase transformation is accompanied by about a 6% *decrease* in volume, whereas the reverse phase transformation is accompanied by about a 6% *increase* in volume, which provides further confirmation for the reversible phase transformation mechanism. The measured densities of nano-grained c-Y_2O_3 and m-Y_2O_3 phases are comparable to the theoretical densities of 5.03 and 5.41 g/cm^3, respectively.

In addition, hardness measurements show that the forward phase transformation from cubic-to-monoclinic increases hardness by ~35%, whereas the reverse phase transformation from monoclinic-to-cubic reduces hardness by ~15%. Hence, the increase in hardness of c-Y_2O_3 by reducing its grain size from 300 μm to 0.1 μm is about 20%. This is indicative of a trend of increase in hardness with decrease in grain size for c-Y_2O_3, perhaps obeying the familiar Hall-Petch relationship [12,13]. On-going research should resolve this issue.

The effect of holding time at 1000°C/8GPa on the forward transformation from c-Y_2O_3 to m-Y_2O_3 is shown in **Figure 3**. Even after 1 min. holding time, complete transformation to the monoclinic phase is accomplished. This suggests that the forward transformation is a displacive transformation. High resolution TEM provides additional evidence for such behavior, since each of the original micro-grained c-Y_2O_3 phase is broken up into nano-domains of the m-Y_2O_3 phase with no clear evidence for grain boundaries.

When holding time increased to 240 min, two additional strong diffraction peaks show their appearance, **Figure 3 (c)**; also detectable after 60 min. This is due to the formation of a new surface-localized phase, which is oxygen deficient and has a cubic symmetry. Apparently, this phase is formed by reaction of the sample with the resistively-heated graphite crucible during HPS. The new surface phase exhibit columnar morphology with nearly square cross section of 2 μm side width and 20 μm length, **Figure 4**. The observation of heavily deformed, and apparently bent, columnar grains suggest that these micro columns have experienced a load higher than the critical load and therefore buckling and plastic deformation occurred without fracture. Plastic deformation in this columnar-grained structure can only occur due to superplastic compressive flow occurs at 8GPa/1000°C, similar to what it has been observed recently in MgB$_2$ [14]. This phenomenon never been observed in Y$_2$O$_3$ system under these conditions, and it is being studied currently in details to understand the role of different parameters including carbon content. Nevertheless, this layer can easily be removed when needed by polishing, since it is not nearly as hard as the underlying matrix material.

SEM observations from both monoclinic (**Figure 5**), and cubic (**Figure 6**) phases show fine features (<200 nm), in agreement with XRD observations. Moreover, detailed TEM analyses of the cubic phase show fine crystallite sizes (20-100 nm) as can be observed from the dark field image and diffraction pattern (**Figure 7 (a)**). All reflections from diffraction pattern matches well with cubic phase with some trace of monoclinic phase as can be seen from weak diffraction from the (310) plane indicated by an arrow. A rough estimate of the retained monoclinic phase is ~5%. HRTEM analysis of one of the crystallites (**Figure 7 (b)**), which was well aligned with electron beam with [$\bar{1}11$] zone axis reflects the cubic symmetry and has size of ~ 100 nm.

From these observations, it is concluded that a much finer nano-grained c-Y$_2$O$_3$ structure can be obtained by reducing transformation temperature and increasing holding time under pressure. Note that without the application of pressure, due to the large internal strains developed during the reverse phase transformation, the sample experiences extensive cracking.

SUMMARY

A pressure-induced reversible phase transformation in polycrystalline Y$_2$O$_3$ accompanied by reduction in grain size from micro- to nano-scale dimensions has been demonstrated. Further work is underway to establish the optimal processing parameters (pressure-temperature-time) to achieve the finest possible nano-grain size in the fully transformed ceramic. Tests are also being made to determine the applicability of this new processing methodology to other technologically important oxide and non-oxide ceramics. An evidence of superplastic compression behavior has been observed in oxygen deficient Y$_2$O$_3$ surface columnar grains.

ACKNOWLEDGMENT
This research is supported by a grant from the Office of Naval Research (ONR).

REFERENCES

1. "High pressure and low temperature sintering of bulk nanocrystalline TiO$_2$," S.-C. Liao, K.D. Pae and W.E. Mayo, Mater. Sci. Eng. A, A204, 152-59 (1995).

2. "The effect of high pressure on phase transformation of nanocrystalline TiO$_2$ during hot-pressing, S.-C. Liao, K.D. Pae and W.E. Mayo, Nanostruct. Mater., 5, 319-23 (1995).

3. "Retention of nanoscale grain size in bulk sintered materials via a pressure-induced phase transformation," S.-C. Liao, K.D. Pae and W.E. Mayo, Nanostruct. Mater., 8-645-56 (1997).

4. "Theory of high pressure/low temperature sintering of bulk nanocrystalline TiO$_2$," S.-C. Liao, W.E. Mayo and K.D. Pae, Acta Mater., 45 [10], 1063-79 (1998).

5. "High pressure/low temperature sintering of nanocrystalline Al$_2$O$_3$," S.-C. Liao, Y.-J. Chen, B.H. Kear and W.E. Mayo, Nanostruct. Mater. 10, 1063-79 (1998).

6. "Breaking the nanograin barrier in sintered ceramics," J. Colaizzi, B.H. Kear and W.E. Mayo, Proc. First Intn'l Conf. on Advanced Materials, eds. D.L. Zhang, K.L. Pickering and X.Y. Xiong: Inst. of Materials Engineering, Australasia, 2000.

7. "On the processing of nanocrystalline and nanocomposite ceramics," B.H. Kear, J. Colaizzi, W.E. Mayo and S.-C. Liao, Scripta Mater. 44, 2065, 2001.

8. "Dense nanoscale single- and multi-phase ceramics sintered by transformation-assisted consolidation," J. Colaizzi, W.E. Mayo, B.H. Kear and S.-C. Liao, Intn'l J. of Powder Metallurgy, 37, 45-54, 2001.

9. "Processing and properties of $ZrO_2(3Y_2O_3)$-Al_2O_3 nanocomposites," R.K. Sadangi, V. Shukla and B.H. Kear, Intn'l J. Refract. Metals & Hard Matls., in press.

10. "Submicron-grained transparent yttria composites," B.H. Kear, R. Sadangi, V. Shukla, T. Stefanik and R. Gentilman, Proceedings of SPIE Conf. on Window and Dome Technologies and Materials IX, March 2005, Orlando, FL, p. 227.

11. "High pressure-high temperature device for making diamond materials," O.A. Voronov, G.S. Tompa and B.H. Kear, Diamond Materials VII, vol. 2001-25, p. 264-271, The Electrochemical Society, 2001.

12. "The deformation and ageing of mid steel: III Discussion of results", Hall EO. Proc Phys Soc Lond B;64:747–53, 1951

13. "The cleavage of polycrystals", Petch NJ., J. Iron Steel Inst;174:25–8, 1953

14. "Superplastic compression flow in MgB2", John D. DeFouw, David C. Dunand, Acta Materialia, 57, 4745-4750, 2009

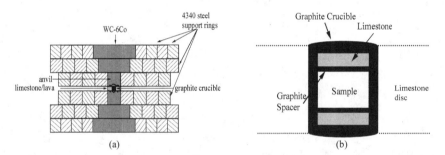

(a)

(b)

Figure 1 - Schematic of high pressure/high temperature cell design, showing (a) arrangement of WC/Co anvils and 4340 steel support rings, and (b) resistively-heated graphite crucible and insulating limestone discs.

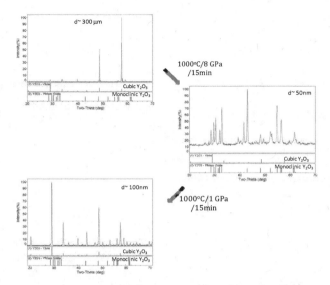

Figure 2 - XRD spectra showing a significant reduction in grain size (300 µm to 0.1 µm) for Y$_2$O$_3$, when subjected to a pressure-induced reversible phase transformation.

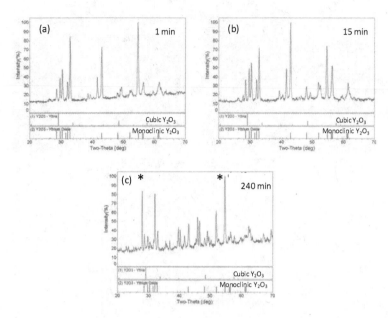

Figure 3 - XRD spectra showing the effect of holding time on the forward transformation from c-Y$_2$O$_3$ to m-Y$_2$O$_3$ at 1000°C /8GPa: (a) 1 min, (b) 15 min, and (c) 240 min.

Figure 4 – SEM micrograph of localized surface columnar grains appeared after the forward transformation from c-Y$_2$O$_3$ to m-Y$_2$O$_3$ at 1000°C /8GPa/ 360 min.

(a)   (b)

Figure 5 - Low (a) and high (b) magnification SEM micrographs of fracture surfaces of m-Y_2O_3 sample, after forward transformation at 1000°C/8GPa/15 minutes.

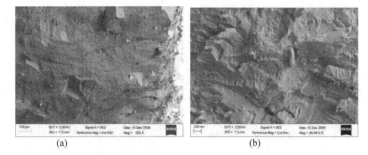

(a)   (b)

Figure 6 – Low (a) and high (b) magnification SEM micrographs of fracture surfaces of c-Y_2O_3 after reversible transformation at 1000°C /1GPa/15min.

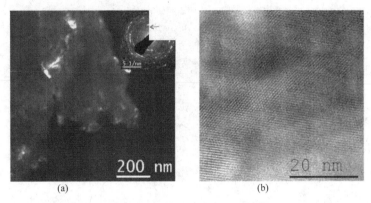

(a)   (b)

Figure 7 – Dark field image (a) and high resolution TEM micrograph (b) of c-Y₂O₃ after reversible transformation at 1000°C /1GPa/15min. The inset in (a) is a diffraction pattern of the imaged region showing a weak signal of (310) reflection from the monoclinic phase (indicated by arrow).

SYNTHESIS OF NANO-SIZE TIB₂ POWDERS USING CARBON COATED PRECURSORS

R. Duddukuri and R. Koc
Mechanical Engineering and Energy processes Department
Southern Illinois University-Carbondale
Carbondale, IL 62901

J. Mawdsley and D. Carter [*]
Argonne National Laboratory
9700 S. Cass Avenue
Argonne, IL 60439

ABSTRACT

This research deals with the synthesis and characterization of titanium diboride (TiB₂) from novel carbon coated precursors. This work provides information on using different boron sources and their effect on the resulting powders of TiB₂.The process has two steps in which the oxide powders were first coated with carbon by cracking of a hydrocarbon gas, propylene (C_3H_6) and then, mixed with boron carbide and boric acid powders in a stoichiometric ratio. These precursors were treated at temperatures in the range of 1200-1400° C for 2 h in flowing Argon atmosphere to synthesize TiB₂.The process utilizes a carbothermic reduction reaction of novel carbon coated precursor that has potential of producing high-quality powders (sub-micrometer and high purity). Single phase TiB₂ powders produced, were compared with commercially available titanium diboride using X-ray diffraction and Transmission electron microscopy obtained from boron carbide and boric acid containing carbon coated precursor.

INTRODUCTION

The design and manufacture of advanced ceramic materials for applications at high temperature and stress is one of the most challenging tasks of modern engineering. Refractory materials such as borides, nitrides and carbides are the natural candidates for these demanding applications due to their excellent hardness and stability at high temperature. "Titanium Diboride (TiB₂), is a refractory compound of covalent and metallic bonding in the hexagonal structure cause a combination of properties characteristics of both ceramics and metals" [2]. TiB₂ has high melting point (2980° C) and high hardness value (18-27 GPa) respectively. This material also has high resistance to oxidation at elevated temperatures and is stable in metal melts (Al, Cu, Mg, Zn, Fe, and Pb) [1]. In addition its high electrical conductivity (22×10^6 Ω cm), good thermal conductivity (96 W/m/K) and considerable chemical stability makes it ideal for high temperature structural and wear applications [5]. But, "these properties are strongly influenced by the microstructure and therefore by the starting powder and precursor compositions and the processing parameters" [2]. Hence, current applications of TiB₂ appear in impact resistant armor, cutting tools, crucibles and wear resistant coatings. Because of its high temperature properties, it is the best choice for control rod material in nuclear reactors and electrodes in aluminum industry.

[*] This research was supported under contract no: DE-AC02-06CH11357, U.S. Department of Energy and the University of Chicago Argonne, LLC, representing Argonne National Laboratory.

In the present investigation, TiB_2 will be synthesized to be utilized as a filler material for bipolar plates in PEM (Polymer electrolyte membrane) fuel cell. Titanium diboride exhibits properties like excellent resistance to oxygen up to 1000°C, high chemical stability and most importantly good thermal conductivity which makes it an ideal filler material for bipolar plate in PEM fuel cell. Titanium diboride powder has been produced commercially by reducing titanium oxide with either boron oxide and carbon or an alkali metal and boron oxide. The carbothermal synthesis of titanium diboride powder is by far the cheapest process because it uses inexpensive starting materials and the effective process to produce large amount of powder at one time. In the current paper, titanium diboride powders are synthesized by reacting different boron sources with carbon coated titanium oxide. The two boron sources used for the study are boron carbide and boric acid. Also for each mole of TiB_2 produced, the process generates CO gas which will release energy when burnt with oxygen. In reality, Equation (1) & (2) both produce $CO_{(g)}$, however Equation (1) is the final state of Equation (2) if the $CO_{(g)}$ is allowed time to form $CO_{2(g)}$. Therefore, the length of reaction time and temperature determines which reaction takes precedence [5]. The only issue with the carbothermal process is that the particles produced, has large size and irregular shapes and the reason is all the starting materials must be mixed mechanically. Regardless, of the mixing time the mechanical mixing will never have intimate contact between C and TiO_2. Without intimate contact, the initial reaction will take time to begin until a higher temperature is reached. Hence, the particles produced must be milled and crushed to make them submicrometer, causing loss of spherical shape, increasing size distribution, and reducing purity.

$$2TiO_2 + C_{(s)} + B_4C_{(s)} = 2TiB_{2(s)} + 2CO_{2\,(g)} \qquad (1)$$

$$2TiO_{2(s)} + 3C_{(s)} + B_4C_{(s)} = 2TiB_{2(s)} + 4CO_{\,(g)} \qquad (2)$$

$$TiO_2 + 2H_3BO_3 + 5C = TiB_{2(s)} + 3H_2O + 5CO_{\,(g)} \qquad (3)$$

RESULTS AND DISCUSSUIONS
Precursor
Fig.1 shows the XRD of the TiO_2 powder coated with carbon using the process developed by Koc *et al* [3,4]. The starting material used for the synthesis of the TiB_2 powder has a crystalline like structure. X-rays diffract off these crystalline structures and generates the theoretical patterns. The starting material commercial TiO_2 (P-25 Degussa) and the coated precursor were X-rayed and their graphs are shown in the Figure 1.
The coated precursor (TiO_2 + C) consists of titania (TiO_2) with amorphous carbon deposited on its surface. X-rays diffract off the TiO_2 peaks and gets absorbed wherever carbon is present. During the coating process, TiO_2 particles were which present in its anatase and rutile forms were reduced to its intermediate oxides such as $Ti_{0.72}O_2$, $Ti_{0.784}O_2$. This is due to the beginning of reduction process at the precursor preparation step. Despite the partial reduction of TiO_2, there was no indication of carbide formation during the coating process. The XRD in Fig. 1 clearly shows no carbide formation. Hence, it is clear that carbon forms a thick layer around the surface of TiO_2 particle instead of existing as particle. The TEM morphology in Fig.2 significantly shows deposition of carbon as a layer around TiO_2. The coating process allows for intimate contact between TiO_2 and C.

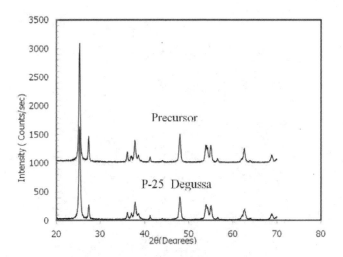

Figure 1: XRD of Carbon coated Precursor

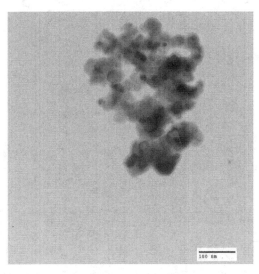

Figure 2: TEM morphology of Coated Titania Precursor

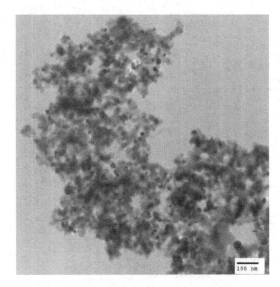

Figure 3: TEM morphology of Coated Titania Precursor

The intimate contact of carbon with TiO$_2$, Ti$_x$O$_y$ (lower oxides), B and CO$_2$ is essential during the reduction reactions for complete conversion of TiO$_2$ and B$_4$C/ TiO$_2$ and H$_3$BO$_3$ into TiB$_2$. Due to the readily availability of carbon around the TiO$_2$ particles there is continuous supply of CO$_{(g)}$, which is required for the reduction of TiO$_2$ to lower oxide form TiO which further reduces to Ti particles. Finally, the reduced the Ti particle reacts with the reduced B particle to form TiB$_2$. In addition, this process helps in producing TiB$_2$ at low temperatures since the reactions proceeds in gas phase because the carbon is a hydrocarbon gas.

Synthesis of Titanium Diboride Powder

Weight loss

The weight loss was the first criteria of characterization after the powders were reacted in the furnace. The weight loss obtained at different formation temperatures are shown in the Table 1. The theoretical weight loss value for Equation 2 is 44.62 % and for the Equation 3 is 73.62 %. The thermodynamic data for these equations was generated from HSC Chemistry 5.1 software (Outo kumpu, Oy, Pori, Finland). Clearly, at 1200°C the reaction did not go to completion due to the lower temperature and lesser amount of reaction time. And at, 1400°C the reaction went to completion consuming all the reactants. Even, 1340°C the reaction almost went to completion and anything less than 1300°C will have lower percentage of weight loss.

Table I: Weight loss % obtained for TiB$_2$ from the Precursor Powder

Temperature (°C)	Percentage weight loss from boron carbide precursor	Percentage weight loss from boric acid precursor
1200°C	24.2	41.9
1340°C	40.5	72.8
1400°C	42.7	73.8

X-Ray Diffraction (XRD)

The XRD of the TiB$_2$ obtained from the boron carbide precursor method is shown in Figure 4. For comparison, the XRD of TiB$_2$ (-325 mesh) from CERAC corporation is also shown. In Figure 4, as the temperature increases from 1200°C-1400°C the intensity of the TiB$_2$ peaks also grows. And, the peaks obtained at 1400°C completely free from impurities/ other phases which indicates TiB$_2$ produced were crystalline pure. Although, the peaks of TiB$_2$ from 1340°C and 1200°C are also free from impurities but these powders might have some excess carbon which remained un-reacted during the process. This reason is supported by the weight loss data obtained in the Table I.

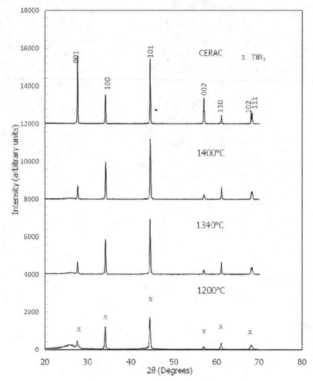

Figure 4: XRD of TiB$_2$ from boron carbide Precursor

Fig. 4(b) illustrates as the temperature increases the formation of TiB$_2$ from boric acid precursor also increases. At 1200°C, the TiB$_2$ peaks have just started to form from the intermediate reaction between TiO and the boron source and produced a weight loss of 42%. While at 1340°C, as the reaction proceeded from 1200°C to this temperature, the intermediate oxide phases were completely eliminated and formed TiB$_2$ with the some free carbon left in the system. A weight loss of 72.8% was recorded at 1340°C which indicates the reaction occurred is closer to completion.

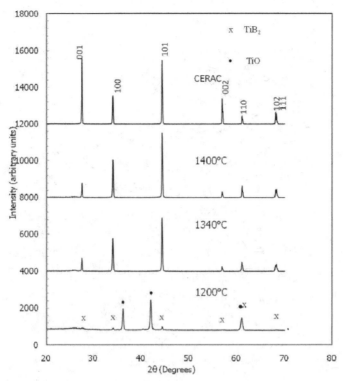

Figure 5: XRD of TiB₂ from boric acid Precursor

Transmission electron microscopy (TEM)

TEM investigations were performed on the TiB₂ powder produced from the boron carbide and boric acid precursor mix for understanding the size distribution, particle shape and degree of agglomeration produced. The TEM micrograph for the TiB₂ obtained from boron carbide precursor mix at 1400°C is shown in Figure 6. It shows the particles are of fairly uniform size accompanied by some odd large particles which are of sub-micrometer range. And, the fine particles were of the ranging from ~ 0.5-0.6 μm. Further, milling (15-20 min) can be performed to break the large size particles into much more fine size one with effect of increasing the surface area of the TiB₂ powder.

Fig.7 is the TEM morphology of TiB₂ obtained from boric acid precursor at 1400°C. The particle size distribution is wide comparatively from the TiB₂ obtained from the boron carbide precursor. And, the reason is boron source used, which has large particle size distribution. In Fig. 7 that the large size TiB₂ particles (2-3 μm) are accompanied by very fine size particles which are sub-micrometer in size. Hence, the produced product is subjected to milling in a plastic vial for 15 minutes using methacrylate balls (plastic) to break the large TiB₂ particles.

TEM micrograph of TiB$_2$ from CERAC Corporation is also shown in figure 8, for comparing the particle size distribution which is ranging in micrometer.

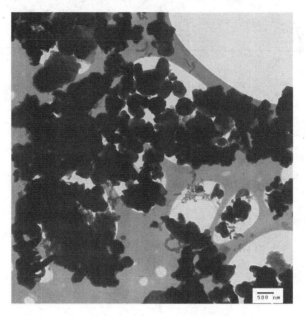

Figure 6: TEM morphology of TiB$_2$ obtained from boron carbide precursor at 1400°C

Figure 7: TEM morphology of TiB$_2$ obtained from boric acid precursor at 1400°C

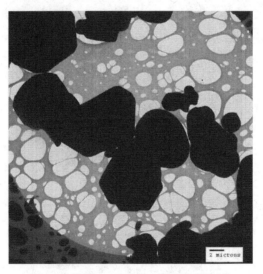

Figure 8: TEM morphology of TiB$_2$ – CERAC

Scanning electron microscopy (SEM)

Figure 9: TiB$_2$ produced from boron carbide   Figure 10: TiB$_2$ produced from boric acid

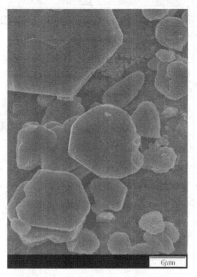

Figure 11: TiB$_2$ obtained from CERAC
Corporation

The Figure 9, 10 and 11 are SEM morphologies for TiB$_2$ obtained from boron carbide precursor, boric acid precursor and finally TiB$_2$ from the CERAC. These morphologies indicate the larger particles were spherical shaped and demonstrate wide particle size distribution.

CONCLUSIONS

The optimal temperature for the production of TiB$_2$ via carbothermal reduction process from the boron carbide and the boric acid were determined to be as 1400°C using the carbon coated precursors. And, the TiB$_2$ produced from the carbon coated method gave fine size particles (0.5-0.6μm) with spherical geometry. Single phase TiB$_2$ powders were obtained with high purity. Also, single phase TiB$_2$ powder were obtained at as low as 1200°C which is the superior result compared to any other carbothermal reduction process. The powders produced are loosely agglomerated requiring minor milling of the powders. The high purity TiB$_2$ was produced at comparatively low temperature, because the carbon source was a hydrocarbon gas. As a result, the process yielded complete reaction of the reactants producing TiB$_2$ powder with required specifications of very fine size and uniform shaped particles. Another, advantage attained was by the elimination of extensive milling process required for precursor mixing, leading to production of powder at a very low cost. Finally, the TiB$_2$ synthesized from the carbon coated precursor gave superior results as compared with the commercial CERAC TiB$_2$ powder.

ACKNOWLEDGEMENTS

This Research performed at Southern Illinois University at Carbondale, sponsored by the U.S. Department of Energy through Argonne National Laboratory Contract No: DE-AC0206CH11357.

REFERENCES

[1] D. Radev, D. Klissurski, "Mechanochemical Synthesis and SHS of diborides of Titanium and Zirconium", Journal of Materials synthesis and Processing, V. 9, No. 3 (2001).

[2] Y. G. Gogtsi, R. A. Andrievski, "Materials Science of carbides, nitrides and borides", Proceedings of NATO Advanced Study Institute on Materials Science of carbides, nitrides and borides, V. 68, pp. 273 (1998).

[3] Koc and Glatzmaier, "Process for Synthesizing Titanium Carbide, Titanium Nitride and Titanium Carbonitride", U.S. Patent #5,417,952 (1995).

[4] Glatzmaier and Koc, "Method for Silicon Carbide Production by reacting Silica with Hydrogen gas", U.S. #5,324,494 (1994).

[5] Thesis by David Bodie, Southern Illinois University at Carbondale, "New Method for the Production of Submicrometer Titanium diboride" (1999).

HIGH TEMPERATURE DIFFRACTION STUDY OF IN-SITU CRYSTALLIZATION OF TiO$_2$ PHOTOCATALYSTS

I.M. Low[1], W.K. Pang[1], V. De La Prida[2], V. Vega[2], J.A. Kimpton[3] and M. Ionescu[4]
[1]Centre for Materials Research, Curtin University, GPO Box U1987, Perth, WA 6845, Australia
[2]Department of Physics, University of Oviedo, Spain
[3]The Australian Synchrotron, 800 Blackburn Road, Clayton, VIC 3168, Australia
[4]Australian Nuclear Science and Technology Organisation, Sydney NSW 2234, Australia

ABSTRACT

By virtue of a high surface-to-volume ratio, very high photocatalytic activity has already been demonstrated for nanostructured TiO$_2$ with various morphologies such as nano-powders, nano-rods, nano-wires, nano-fibers, nano-belts, nano-tubes, and thin films. Although a lot of progress has been achieved in these forms of TiO$_2$, the poor recuperability and reutilization limitation for nano-powders and processing difficulty for nano-fibres or nano-tubes are still challenges for their commercial applications. In this paper, the synthesis and characterisation of in-situ crystallization of TiO$_2$ nanotubes from the precursor of as-anodized amorphous TiO$_2$ nanotubes after annealing at temperatures up to 900 °C are described. The resultant microstructures and composition depth profiles are discussed in terms of ion-beam analysis and grazing-incidence synchrotron radiation diffraction.

INTRODUCTION

TiO$_2$ is currently used in a number of fascinating applications, including as photocatalysts. In particular, there is great interest in using TiO$_2$ nanotubes because of the added surface areas available. Due to its high photoactivity, photodurability, chemical and biological inertness, mechanical robustness and low cost, nanostructured semiconducting TiO$_2$ has attracted much more attention for its potential applications in diverse fields such as photocatalysis of pollutant,[1] photo-splitting of water[2-4] and transparent conducting electrodes for dye-sensitized solar cells.[5] Since photocatalytic reactions mainly take place on the surface of the catalyst, a high surface-to-volume ratio is of great significance for increasing the decomposition rate. Very high photocatalytic activity has already been demonstrated for nanostructured TiO$_2$ with various morphologies such as nano-powders,[1-5] nano-rods,[6] nano-wires,[7] nano-fibers,[8,9] nano-belts,[10] nano-tubes,[11] thin films,[12] and porous nanostructures.[13-15] Although a lot of progress has been achieved in these forms of TiO$_2$, the poor recuperability and reutilization limitation for nano-powders and processing difficulty for nano-fibres or nano-tubes are still challenges for their commercial applications.

Hitherto, various processes of materials synthesis have been developed for nanostructured TiO$_2$ and these include the sol-gel method,[16] micelles and inverse micelles method,[17] hydrothermal method,[18] and the solvothermal method.[19] In this paper, we describe a direct oxidation of anodized titanium foils to grow a thin surface coating of nano-structured TiO$_2$. Using indirect oxidation methods, nanorod arrays have been prepared on the surface of Ti substrates by oxidizing titanium with $(CH_3)_2CO$,[20] H_2O_2 solution,[21] and KOH.[22] However, few literature sources have reported the growth of aligned TiO$_2$ nanostructures on the substrate by electrochemical anodization of Ti metal and subsequent annealing in air. The growth of aligned TiO$_2$ nanostructures on the substrates may have potential for photoelectrochemical and optoelectronic applications.

177

EXPERIMENTAL METHODS
Sample Preparation
    Ti foils (99.6 % purity) with dimensions of $10 \times 10 \times 0.1$ mm^3 were used for the anodizing to produce self-organized and well-aligned TiO$_2$ nano-tube arrays. The process of potentiostatic anodization was performed in a standard two-electrode electrochemical cell, with Ti as the working electrode and platinum as the counter electrode. Prior to anodization, Ti-foils were degreased by sonicating in ethanol, isopropanol and acetone for 5 minutes each, followed by rinsing with deionised water, and then drying using nitrogen stream. After drying, the foils were exposed to the electrolyte which consists of 100 ml of Ethylene glycol + 2.04 ml of water + 0.34 g of NH$_4$F. The electrolyte's pH was kept constant at pH = 6, and its temperature was kept at about 25 $^\circ$C. The anodization process was performed under an applied voltage of 60 V for 20 h. After that, the resulting TiO$_2$ structures were then rinsed in ethanol, immersed in Hexamethyl-dililazane (HMDS) and dried in air. Some of the samples were subsequently annealed at 400 $^\circ$C in air for 2 hours, whereas as-anodized Ti-foils were also produced for the in-situ study of TiO$_2$ formation at elevated temperature using synchrotron radiation diffraction.
    The morphologies of the anodized Ti samples were characterized using a field emission scanning electron microscope (FESEM SUPRA 35VP ZEISS) operating at working distances of 5 mm with an accelerating voltage of 5 kV.

In Situ High-Temperature Synchrotron Radiation Diffraction
    In this study, the in-situ oxidation behaviour of as-anodized Ti-foils was characterised using high-temperature synchrotron radiation diffraction (SRD) up to 900 $^\circ$C in air. All measurements were conducted at the Australian Synchrotron using the Powder Diffraction beamline in conjunction with an Anton Parr HTK20 furnace and the Mython II microstrip detector. The SRD data were collected at an incident angle of 3° and wavelength of 0.11267 nm.
    The phase transitions or structural changes were simultaneously recorded as SRD patterns. These patterns were acquired at high temperatures over an angular range of 100° in 2 . The SRD patterns were acquired in steps of 100 $^\circ$C from 200 $^\circ$C to 400 $^\circ$C and thereafter every 50 $^\circ$C from 200 $^\circ$C to 400 $^\circ$C. Figure 1 shows the heating protocol of the heating schedule during the experiment. The collected XRD data were analysed and refined using the Rietveld method to compute the relative phase abundances of oxides formed at each temperature. The mean crystallite size (L) of TiO$_2$ was calculated from (101) reflections using the Scherrer equation:[23]

$$L = \frac{K\lambda}{\beta \cos\theta} \qquad (1)$$

where K is the shape factor, λ is the x-ray wavelength, β is the line broadening at half the maximum intensity (FWHM) in radians, and θ is the Bragg angle.

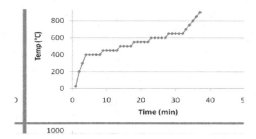

Fig. 1: The heating protocol used for the high-temperature SRD experiment.

Grazing-Incidence Synchrotron Radiation Diffraction

Grazing-incidence synchrotron radiation diffraction (GISRD) is a widely used method for depth profiling thin-layer, multi-layer, and graded materials. In GISRD, the intensity-weighted penetration depth (d) of x-ray penetration into the sample depends on the value of incidence angle (α) and the energy of the x-ray beam (E). The penetration depth is described by:

$$d = 2\alpha/\mu \qquad (2)$$

where μ is the linear attenuation coefficient of the material. SRD depth profiles of anodized and annealed Ti-foils were measured at the Australian Synchrotron on the Powder Diffraction beamline. Imaging plates were used to record the diffraction patterns at incidence angles of 0.3, 0.5, 1.0, 2.0 and 3.0° using a fixed wavelength of 0.13772 nm.

Ion-Beam Analysis

The near surface and oxygen depth profiling study of anodized and annealed Ti-foil samples were conducted by RBS at the Australian Nuclear Science & Technology Organisation (ANSTO) using 1.8MeV He^{1+} ions on the 2MV tandem accelerator. This study allowed the measurement of total thickness and the depth distribution of Ti and O of the process-influenced layer.

RESULTS AND DISCUSSION
In-Situ Formation of TiO$_2$ at Elevated Temperature

Figure 2 shows the synchrotron radiation diffraction plots of as-anodized Ti-metal before and after thermal annealing in air at 20 C – 900 C. The TiO$_2$ nanotube arrays before thermal treatment were amorphous and remained so up to 300 C which eventually crystallized to anatase after thermal treatment at 400 C. However, further heating above 400 C did not cause anatase to transform to rutile even up to 900 C. The complete absence of anatase to rutile transformation at 900 C observed here is very unique and has not been previously reported in the literature. It is well-known that anatase undergoes a phase transformation to rutile at ~500-550 C.[24-26] However, it remains a mystery why the anatase formed in this study was remarkably stable and resistant to phase transformation.

Although anatase remained very stable from 400 to 900 C, there was a distinct narrowing and sharpening in the (101) peak, resulting in a corresponding decrease in the values of full-width half-maximum (FWHM) as the temperature increased. From the Scherrer equation,[23] a decrease in the value of FWHM implies an increase in the mean crystallite size for anatase. Figure 3 shows the influence of annealing temperature on FWHM and crystallite size for anatase. The crystallite size of TiO$_2$ was about 55 nm at 400 C and increased gradually with temperature to just over 70 nm at 900 C. Another feature worth-noting in Fig. 2 is the lack of line-shifts in the peaks as the temperature increases due to thermal expansion. This may imply that TiO$_2$ nanotubes have a very low coefficient of thermal expansion.

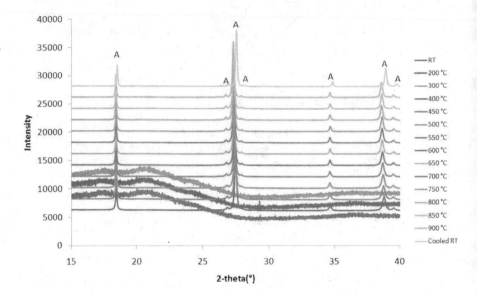

Fig. 2: High-temperature synchrotron radiation diffraction plots showing the in-situ crystallization of anatase from the amorphous TiO$_2$ in the temperature range 20 – 900 °C.

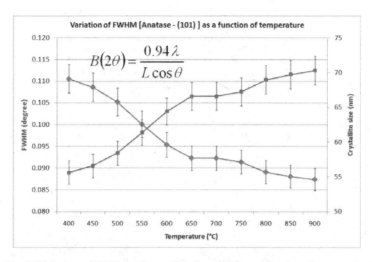

Fig. 3: Variations of FWHM and crystallite size of TiO$_2$ as a function of temperature.

Microstructures and Formation Mechanisms of Nanostructured TiO$_2$

After anodizing in the electrolyte of NH$_4$F / ethylene glycol at 25 °C for 20 h, a homogeneous white layer of TiO$_2$ nanotubes formed on the Ti-foil. Fig. 4(a) shows the FESEM image of the microstructure of amorphous TiO$_2$ nano-stripes on the as-anodized sample. The width of the "stripe" was ~ 80-85 nm while that of length was ~ 3-4 μm. It is unclear why the amorphous TiO$_2$ displayed as nano-stripes rather than nanotubes, although damage or collapse of nanotubes due to prior handling could be a cause. Another probable cause is the drying process itself.[27] Due to the presence of water and F$^-$ ions in the nanostructured tubes, they tend to bend and break during the drying process. Hence, DMSO is usually used to minimize the problem of breakage problems

However, upon annealing at 400 °C for 2 h, vertically oriented and highly ordered arrays of anatase nanotubes with diameter of ~80 nm formed (Fig. 4(b)) which concurs well with other studies on anodized TiO$_2$.[28-30] Note that some nanostripes can still be observed on the left-hand side of Fig. 4(b). The EDS analysis of nanotubes in Spectrum 1 of Fig. 4(c) indicated 31.04% Ti and 68.96% O, thus confirming the presence of TiO$_2$ in the sample.

(a)

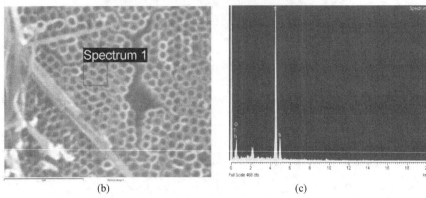

(b)   (c)

Fig. 4: Scanning electron micrographs showing the microstructure of (a) as-anodized TiO_2, (b) anodized and annealed TiO_2 at 400 °C for 2 h, and (c) EDS analysis of composition in Spectrum 1.

Three simultaneously occurring processes can be ascribed to the formation of the nanotube arrays in fluoride containing electrolyte.[31] The first involves the field-assisted oxidation of Ti metal to form TiO_2, which is followed by field-assisted dissolution of Ti metal ions in the electrolyte, and finally the chemical dissolution of Ti and TiO_2 in the presence of hydrogen and fluoride ions. Plausible pathways of chemical reaction occurring during anodization are as follows:

$$Ti\,(s) \rightarrow Ti^{4+}\,(aq) + 4e^-$$
$$Ti^{4+}\,(aq) + 4F^-\,(aq) \rightarrow TiF_4\,(aq)$$
$$TiF_4\,(aq) + 2H_2O\,(aq) \rightarrow TiO_2\,(s) + 4HF\,(aq)$$

Composition Depth-Profiles of TiO_2 Nanotubes

Figure 5 shows the grazing-incidence synchrotron radiation diffraction plots of an anodized and annealed TiO_2 sample at 400 °C for 2 h. The intensity of the anatase peaks at $2\theta = 40 - 55$ clearly changes with an increase in the grazing angle from 0.3 to 3.0°. This change in peak intensity with grazing-angles suggests the presence of composition gradation at the near-surface of the sample.[25,26] The existence of composition gradation with the sample at the near-surface has been verified by the ion-beam analysis result shown in Figure 6.

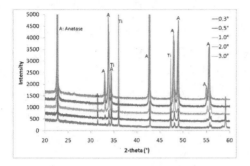

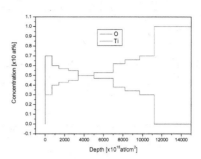

Fig. 5: GISRD plots of anodized TiO_2 sample annealed at 400 °C for 2 h.

Fig. 6: Ion-beam analysis of anodized and annealed TiO_2 sample.

CONCLUSIONS

The in-situ crystallization of anatase on an as-anodized Ti-foil in the temperature range has been investigated using synchrotron radiation diffraction. The as-anodized TiO_2 was amorphous but crystallized to anatase at 400 °C. The anatase formed was very stable and did not transform to rutile even after annealing at 900 °C. However, increasing the temperature from 400 to 900 °C caused the FWHM of (101) peaks to decrease, resulting in a concomitant coarsening in crystallite size from ~55 nm to 70 nm. The surface of annealed sample exhibited well-aligned, uniform TiO_2 nanotubes with an average diameter of ~80 nm and wall thickness of ~10 nm. GISRD and ion-beam analysis confirmed the existence of composition gradation within the annealed TiO_2 sample at the near-surface.

ACKNOWLEDGMENTS

This work was supported by fundings from the Australian Synchrotron (AS102/PD/1654 & AS102/PD/2494), the Australian Institute of Nuclear Science and Engineering (AINGRA10127) and Spanish MAT2010-20798-C05-04. We thank Ms E. Miller of Curtin Centre for Materials Research for assistance with the SEM work. The scientific support from the University of Oviedo SCT's, particularly to the Nanoporous Membranes Unit, is also acknowledged.

REFERENCES

[1]M.R. Hoffmann, S.T. Martin, W. Choi and D.W. Bahnemann, Environmental Applications of Semiconductor Photocatalysis, *Chem. Rev.* **95**, 69-96 (1995).

[2]A. Fujishma, K. Honda, Electrochemical Photolysis of Water at a Semiconductor Electrode, *Nature* **238**, 37 (1972).

[3]Y.B.Liu, B.X. Zhou, J. Bai, J.H. Li and X.L. Zhang, Efficient Photochemical Water Splitting and Organic Pollutant Degradation by Highly Ordered TiO_2 Nanopore Arrays, *Appl. Catal.* B **89**, 142-148 (2009)

[4]E. Indrea, S. Dreve, T.D. Silipas, G. Mihailescu, V. Danciu, V. Cosoveanu, A. Nicoara, L.E. Muresan, E.J. Popovici, V. Popescu, H.I. Nascu and R. Tetean, Nanocrystalline Semiconductor Materials for Solar Water-Splitting, *J. Alloys Compd.* **483**, 445-449 (2009).

[5]B. O'Regan, M. Grätzel, A Low-Cost, High-Efficiency Solar Cell Based on Dye-Sensitized Colloidal TiO$_2$ Films, *Nature* **353**, 737-740 (1991).

[6]H. Cheng, J. Ma, Z. Zhao and L. Qi, Hydrothermal Preparation of Uniform Nanosize Rutile and Anatase Particles, *Chem. Mater.* **7**, 657-671 (1995).

[7]Z. Miao, Xu D, Ouyang J, Guo G, Zhao X, Tang Yet al., Electrochemically Induced Sol−Gel Preparation of Single-Crystalline TiO$_2$ Nanowires, *Nano Lett.* **2**, 717-720 (2002).

[8]Y.G. Guo, J. Hu, H. Liang, L. Wan and C. Bai, TiO$_2$-Based Composite Nanotube Arrays Prepared via Layer-by-Layer Assembly, *Adv. Funct. Mater.* **15**, 196-202 (2005).

[9]Y. Zhang, Y. Gao, X.H. Xia, Q.R. Deng, M.L. Guo, L. Wan, G. Shao, Structural Engineering of Thin Films of Vertically Aligned TiO$_2$ Nanorods, *Mater. Lett.* **14**, 1614-1617 (2010).

[10]Y.M. Wang, G. Du, H. Liu, D. Liu, S. Qin, N. Wang, C. Hu, X. Tao, J. Jiao, J. Wang and Z.L. Wang, Nanostructured Sheets of TiO$_2$ Nano-Belts for Gas Sensing and Antibacterial Applications, *Adv. Func. Mater.* **18**, 1131-1137 (2008).

[11]Z.R. Tian, J.A. Voigt, J. Liu, B. Mckenzie, and H. Xu, Large Oriented Arrays and Continuous Films of TiO$_2$ Based Nanotubes, *J. Am. Chem. Soc.* **125**, 12384-12385 (2003).

[12]R. Asahi, T. Morikawa, T. Ohwaki, K. Aoki and Y. Taga, Visible-Light Photocatalysis in Nitrogen-Doped Titanium Oxides, *Science* **293**, 269-271 (2001)

[13]P.D. Yang, D. Y. Zhao, D. Margolese, B.F. Chmelka and G.D. Stucky, Generalized Syntheses of Large-Pore Mesoporous Metal Oxides with Semicrystalline Frameworks, *Nature* **396**, 152-155 (1998).

[14]U.M. Patil, K.V. Gurav, O.S. Joo and C.D. Lokhande, Synthesis of Photosensitive Nanograined TiO$_2$ Thin Films by SILAR Method, *J. Alloys Compd.* **478**, 711-715 (2009).

[15]Y. Shen, J. Tao, F. Gu, L. Huang, J. Bao, J. Zhang and N. Dai, Preparation and Photoelectric Properties of Ordered Mesoporous Titania Thin Films. *J. Alloys Compd.* **474**, 326-329 (2009).

[16]J. Tang, F. Redl, Y. Zhu, T. Siegrist, L.E. Brus, and M.L. Steigerwald, An Organometallic Synthesis of TiO$_2$ Nanoparticles, *Nano Lett.* **5**, 543-548 (2005).

[17]G.L. Li, and G. H. Wang, Synthesis of nanometer-sized TiO$_2$ Particles by a Microemulsion Method, *Nanostruct. Mater.* **11**, 663-668 (1999).

[18]Q. Huang, L. Gao, A Simple Route for the Synthesis of Rutile TiO$_2$ Nanorods, *Chem. Lett.* **32**, 638-640 (2003).

[19]X. Wang, J. Zhuang, Q. Peng & Y. Li, A General Strategy for Nanocrystal Synthesis, *Nature* **437**, 121-124 (2005).

[20]X. Peng, A. Chen, Aligned TiO$_2$ Nanorod Arrays Synthesized by Oxidizing Titanium with Acetone, *J. Mater. Chem.* **14**, 2542-2548 (2004).

[21]J.M. Wu, Low-Temperature Preparation of Titania Nanorods through Direct Oxidation of Titanium with Hydrogen Peroxide, *J. Cryst. Growth* **269**, 347-355 (2004).

[22]M. Asadi, M. Attarchi, M. Vahidifar, A. Jafari, Film Formation via Plasma Electrolyte Oxidation of Ti and Ti-5Mo-4V-3Al Alloy in High Alkaline Solutions, *Defect and Diffusion Forum*, **297-301**, 1167-1170 (2010).

[23]B.D. Cullity & S.R. Stock, *Elements of X-Ray Diffraction*, 3rd Ed., Prentice-Hall Inc., pp. 167-171 (2001)

[24]S. Sreekantan, R. Hazaan, Z. Lockman, Photoactivity of Anatase-Rutile TiO$_2$ Nanotubes Formed by Anodization Method, *Thin Solid Films* **518**, 16-21 (2009).

[25]I.M. Low, E. Wren, K.E. Prince & A. Anatacio, Characterisation of Phase Relations and Properties in Air-Oxidised Ti$_3$SiC$_2$. *Mater. Sci. & Eng. A.* **466**, 140-147 (2007).

[26]W.K. Pang, I.M. Low, B.H. O'Connor, K.E. Prince & A. Anatacio, Oxidation Characteristics of Ti$_3$AlC$_2$ over Temperature Range 500-900°C. *Mater. Chem. Phys.* **117**, 384-389 (2009).

[27]X. Meng, T.Y. Lee, H. Chen, D.W. Shin, K.W. Kwon, S.J. Kwon, J.B. Yoo, Fabrication of Free Standing Anodic Titanium Oxide Membranes with Clean Surface Using Recycling Process, *J. Nanosci. & Nanotech.* **10**, 4259–4265, (2010).

[28]V.M. Prida, M. Hernandez-Velez, K R Pirota, A. Menendez and M. Vazquez, Synthesis and Magnetic Properties of Ni Nano-Cylinders in Self-Aligned and Randomly Disordered Grown Titania Nanotubes, *Nanotechnology* **16**, 2696–2702 (2005).

[29]O.K. Varghese, D. Gong, M. Paulose, C.A. Grimes, E.C. Dickeya, Crystallization and High-Temperature Structural Stability of TiO_2 Nanotube Arrays, *J. Mater. Res.* **18**, 156-165 (2003).

[30]Y. Zhang, W. Fu, H. Yang, Q. Qi, Y. Zeng, T. Zhang, R. Ge, G. Zou, Synthesis and Characterization of TiO_2 Nanotubes for Humidity Sensing. *Appl. Surf. Sci.* **254**, 5545–5547 (2008).

[31]K. Mor, O. K. Varghese, M. Paulose, K. Shankar, C. A. Grimes, A Review on Highly Ordered, Vertically Oriented TiO_2 Nanotube arrays: Fabrication, material properties, and solar energy applications, *Solar EnergyMat. Solar Cells.* **90**, 2011-2075 (2006).

Author Index

Author Index